291
Current Topics in Microbiology and Immunology

P. Boquet and E. Lemichez (Eds.)

Bacterial Virulence Factors and Rho GTPases

With 28 Figures and 4 Tables

Springer

Professor Dr. Patrice Boquet
Professor Dr. Emmanuel Lemichez

INSERM U627, IFR 50
Faculty of Medicine
28, Avenue de Valombrose
Nice
France

e-mail: boquet@unice.fr, lemichez@unice.fr

Cover illustration by Pr Pierre Gounon
The book cover shows, by scanning electronic microscopy, a Shigella invading a HeLa cells by a process named "trigger phagocytosis". This process requires the activation of The Rac1 and Cdc42 Rho GTPases by Shigella virulence factors. (By courtesy of Pr Pierre Gounon, Centre de Microscopie de l'Université de Nice Sophia Antipolis, Nice, France.)

ISSN 0070-217X
ISBN 978-3-642-06280-3 e-ISBN 978-3-540-27511-4

Springer is a part of Springer Science+Business Media
springeronline.com

Printed in The Netherlands

Editor: Dr. Rolf Lange, Heidelberg
Desk editor: Anne Clauss, Heidelberg
Production editor: Michael Hübert, Leipzig
Cover design: design & production GmbH, Heidelberg

Printed on acid-free paper 21/3150/YL – 5 4 3 2 1 0

Preface

Pathogenic bacteria for humans and animals have developed sophisticated weapons, termed virulence factors, to ensure their replication and persistence within their hosts. The first role of these virulence factors is to loosen the host defenses against microorganisms represented by innate and adaptative immunities. Small GTPases of the Rho subfamily have been shown to play important functions in these systems. The chapter by B.B. Finlay is thus devoted to the presentation of the Rho GTPases and their effectors in the general strategies of bacterial virulence factors.

Infectious bacteria must first dock on the surface of epithelial cells to colonize their host. Many interactions with molecules involved in epithelial cell–cell interactions or cell substrate recognition with GTPases of the Rho subfamily have been shown to be pivotal for processes such as formation of focal adhesion points and tight and adherens junctions dynamics. Bacterial virulence factors have often hijacked or domesticated the regulatory roles of Rho GTPases to facilitate their binding to the epithelial cell. The chapter by G. Duménil and X. Nassif focuses on how bacteria modulate their adherence on the cell surface via Rho GTPases.

One example of how bacterial virulence factors allow microorganisms to escape host defenses is the triggering of microbe engulfment by epithelial cells using Rho GTPases directly. The chapter by M. Schlumberger and W.-D. Hardt is dedicated to the pathogenic bacteria *Salmonella*, the paradigm of that mechanism.

It is clear now that Rho GTPases are important elements for innate and adaptative immunities. For instance, Rho GTPases are required for the constitution of the immunological synapse. The chapter by M. Deckert et al. therefore deals with activities of Rho GTPases during T cell receptor stimulation. Another important role of Rho GTPases in host defense against infectious microbes is their main function during engulfment of bacteria by polymorphonuclear leukocytes. This is discussed in the chapter by B.A. Diebold and G. Bokoch.

Bacterial toxins have been shown to be the first microbial virulence factors to interfere with Rho GTPases. Toxins either activate or deactivate Rho

GTPases. Furthermore, virulence factors that are *not bona* fide toxins but are introduced by direct injection through bacterial needles manipulate Rho GTPases almost similarly. The knowledge of the mode of action of toxins and toxinlike factors that affect Rho GTPases not only has been essential to understanding the pathogenicity of the producing microbes but the use of these toxins and toxinlike factors as biological probes has led to major breakthroughs in cell biology. The chapter by K. Aktories and I. Just describes the bacterial toxins inhibiting the Rho GTPases. Toxinlike bacterial virulence factors affecting Rho GTPases are described in the chapter by J.T. Baldwin and M.R. Barbieri. The chapter by M. Aepfelbacher et al. is devoted to a virulence factor of the bacterium Yersinia that inhibits Rho GTPases, because discovery of the mode of action of this protein (YopT) has shown similarities between pathogenic mechanisms in bacteria and the plant defense systems. The chapter by P. Munro and E. Lemichez covers toxins known to activate Rho GTPases such as the cytotoxic necrotizing factor of uropathogenic *Escherichia coli* and their implication in bacterial strategies to colonize the host.

We hope that this volume will provide a synthesis on how the various host cellular Rho GTPases activities are manipulated by bacteria to fulfill their virulence and that it will be useful for the scientific community working on cellular microbiology.

P. Boquet, E. Lemichez

List of Contents

List of Contributors

(Their addresses can be found at the beginning of their respective chapters.)

CTMI (2005) 291:1–10

Bacterial Virulence Strategies That Utilize Rho GTPases

B. B. Finlay

Biotechnology Laboratory, University of British Columbia,
Vancouver British Columbia, V6T 1Z3, Canada
bfinlay@interchange.ubc.ca.

Abstract The ability to modify central host cellular functions is a major advantage to many bacterial pathogens that use such strategies as part of their virulence mechanisms. Small GTPases, including Rho GTPases, make particularly attractive targets for pathogens because of their central roles in modulating cellular functions such as cytoskeletal control. Such modifications of these GTPases can include direct chemical modification of the GTPase or interfacing with some of the regulatory elements associated with GTPase control. Pathogens use these alterations in GTPase functions for a variety of functions, including killing the host cell, mediating bacterial uptake into the host cell (invasion), reprogramming actin to form a lesion in host cells underlying adherent bacteria, to mediate intracellular survival by affecting intracellular trafficking, or to provide polymerized actin mechanisms to propel microbes around inside host cells and into adjacent cells. Collectively, these examples represent many key microbial virulence mechanisms that have led to a much deeper understanding of both microbial pathogens and GTPase functions.

1
Introduction

There are approximately 100 pathogenic microbes that cause significant disease in humans, accounting for one-third of all deaths on the planet, in addition to many other pathogens that infect other mammals, animals, and plants. These pathogenic microbes possess many sophisticated virulence strategies that are designed to overcome generally effective host defense mechanisms that defend against the continual exposure to microbes [7]. Generally, these virulence mechanisms target one or more normal host cellular processes, and it is the collective action of these mechanisms that ultimately ends in disease.

The choice of host processes to target are numerous: signaling mechanisms, cell division, host immune response, phagocytosis, epithelial barrier integrity, chemotaxis, phagosome-lysosome fusion, etc.. Most virulence factors target a specific host molecule to mediate these effects. This entails direct contact of the bacterial virulence factor with the appropriate host molecule, which may be on the host cell surface or, in many cases, inside the host cell. Thus bacterial pathogens have developed various strategies to deliver their virulence factors, from binding to the cellular surface, being taken up by normal endocytic routes, and then escaping the membrane bound inclusion (many toxins use this route) to being injected directly into the host cell with specialized type III and type IV secretion systems [11, 14] and then targeting to the appropriate intracellular location. Recent knowledge in cell biology has advanced very rapidly, and much of this progress is due to the use of virulence factors as tools to study normal cellular processes. Indeed, the function of such key cytoskeletal regulators such as Rho, N-WASP, and Arp2/3 were discovered by using bacterial factors that modulate them.

An ideal virulence factor target should be one that controls one or more important cellular processes, and whose manipulation will provide the invading pathogen with a subsequent advantage for surviving and multiplying within the infected host. Thus the more "ideal" the host target, the more examples there are of virulence factors that aim to alter and/or disrupt such a target. One of the most extensively targeted cellular processes is the ability to alter cytoskeletal rearrangements, especially the subset of actin-based processes (as opposed to microtubules or intermediate filaments). The actin-based cytoskeleton has many important roles in eukaryotic cells, including cell motility, phagocytosis, and cell division, and is an essential process to eukaryotic cells. Not surprisingly, many bacterial pathogens go after the "master controls" of the cytoskeleton, the small GTP-binding proteins belonging to the Rho family of GTPases, and have devised many clever ways to activate, inactivate, modulate, and generally manipulate these important cellular regulators, ultimately using these mechanisms as part of their overall virulence strategy.

2 Small GTPases

Although ATP serves as the main energy source within cells, many proteins (called G proteins) can bind and cleave GTP to regulate cellular processes and mechanisms. Generally, G proteins are divided into two groups, with a major family being the small G proteins or small GTPases (20–40 kDa).

Table 1 Small GTPase families and functions

Small GTPase family	General functions
Ras	Regulator of gene expression, cell division and transformation, MAP kinase cascade, cell proliferation and differentiation, apoptosis
Rho (includes Rac and Cdc42)	Modulators of actin cytoskeleton, activation of NADPH oxidase, stress fibers (Rho), lamellipodia and membrane ruffles (Rac), filopodia (Cdc42)
Rab	Intracellular vesicle targeting, docking, and fusion
Sar1/Arf	Vesicle membrane recruitment, including COPII (Sar1) and COPI, AP-1, and AP-3 (Arf)
Ran	Nucleocytoplasmic transport and microtubule organization

This family consists of over 100 members, being found in eukaryotes ranging from yeast to human. These break down into a further five subgroups or families: Ras, Rho, Rab, Sar1/Arf, and Ran [16]. Each family has generalized functions that are essential for normal cellular functions, including signaling, cytoskeletal rearrangements, vesicle targeting, nucleocytoplasmic transport, and microtubule organization (Table 1). (For a very comprehensive review on small GTPases and their functions, see [16].) Currently, four of these families (Ras, Rho, Rab, and Sar1/Arf) are targeted by microbial virulence factors, and, although not documented yet, the Ran family makes an attractive target to disrupt either nuclear transport or microtubule function by microbial pathogens.

All the small GTPases share a common mechanism by which they function, based on the ability to bind and cleave GTP (Fig. 1) [16]. In addition, this mechanism uses several other regulatory proteins, which results in a very finely controlled molecular switch that is used to modulate cell functions. A small GTPase is in the inactive form when it binds GDP. An upstream activation signal stimulates the dissociation of GDP followed by the binding of GTP, which leads to a conformational change enabling the activated GTPase to then bind to and activate downstream effectors (Fig. 1). The intrinsic GTPase activity of the small GTPase then cleaves GTP into GDP, which releases the bound effector, returning the GTPase back to its inactive form. Regulator proteins called GEFs (the guanine nucleotide exchange factors) assist the rate-limiting step of the GDP/GTP exchange by facilitating release of bound GDP followed by subsequent GTP binding, which displaces GEF. Thus GEFs

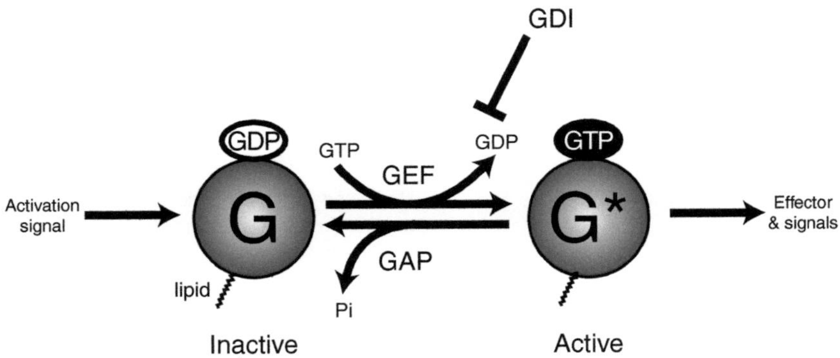

Fig. 1 Generalized mechanism of small GTPase regulation and activity. See text for details

"activate" small GTPases. Another regulator, called GAP (GTPase activating protein) stimulates the GTPase activity further (Fig. 1). Members of the Rab and Rac/Rho/Cdc42 family use an additional regulator called GDI (GDP dissociation inhibitor), which inhibits the dissociation of GDP from the small G protein, keeping the G protein in its (inactive) GDP-bound state. In addition to regulatory molecules that modulate activity, small GTPases are also covalently modified by lipids at their COOH termini, including farnesylation, geranylgeranylation, and prenylation (see [16] for details). Inhibitors that block these lipid modifications block G protein migration to the membrane, which inhibits activity. Collectively, all these regulators and modifications tightly control the activity of small G proteins, allowing the cell to rapidly turn on and off these key elements of cellular function. Not surprisingly, many microbial pathogens have also realized the power of manipulating small GTPases and their regulatory elements and target many G proteins, GAPs, GEFs, and GDIs.

The Rho family of small G proteins contains several family members, including several Rac and Rho members, plus Cdc42. These GTPases are generally thought to control actin cytoskeleton-based functions [10]. There are also numerous effectors upstream that trigger their activity (see [16] for details). In general, Rho family members control stress fibers in cells, which are long, extended bundles of actin that are easily visible along the basolateral plane of cultured mammalian cells. Rac family members regulate membrane ruffles and lamellipodia that occur at active areas of mammalian cell surfaces. Cdc42 binds N-WASP, which then activates the actin polymerization machinery Arp2/3, which mediates filopodia (fingers at the cell surface) formation, as well as other actin-based functions.

3
Small GTPases and Bacterial Virulence Strategies

As discussed above and seen throughout this volume, small Rho GTPases are utilized extensively by bacterial pathogens to usurp normal cellular processes as part of virulence. This is presumably because of their central role in cellular functions and their many diverse downstream effects. These virulence factors generally fall into two categories: toxins that bind to cellular surfaces and can be internalized into host cells and effector proteins that are injected via type III and type IV secretion systems directly into host cells , where they can then target appropriate G proteins. In general, the toxins kill the host cells (thus their name), whereas effectors modulate cellular effects that benefit the pathogen, such as mediating bacterial uptake into host cells (invasion). Each of these processes is discussed in extensive detail throughout the remainder of this volume, and thus the details of each of the processes will not be reiterated here. Instead, a more general overview of the processes as they relate to bacterial virulence is presented in the context of bacterial virulence strategies.

Perhaps the most daunting obstacle a pathogen faces when encountering a potential host is the epithelial barrier. At some points on the body such as skin, this barrier is so impermeable the only way through it is via a break (cut) or at the base of hair follicle sites. However, at other sites in the body such as most mucosal surfaces, the barrier must be more permeable, as the body needs to shunt nutrients and cells such as phagocytes back and forth. The tight seal between epithelial cells (tight junctions or zona occludens) is supported by a band of actin running around the peripheral apical surface of each cell. Because it is made of actin, small Rho GTPases play a key role in maintaining its integrity (and thus cell polarity). Not surprisingly, bacterial pathogens have developed various ways to disrupt this barrier to gain access to deeper tissue or cause diarrhea. Thus Rho GTPases make attractive targets for pathogens that require the breakdown of epithelial barriers. Such breakdown may inhibit the normal immune sampling mechanisms by disrupting the normal cytology, disrupting chemotactic recruitment of macrophages and neutrophils, disrupting normal nutrient uptake, and altering luminal fluid production. It is thought that triggering diarrhea enhances the removal of normal flora that normally compete with an incoming pathogen for limited nutrient supplies. Additionally, enhanced fluid secretion increases the spread of a diarrheagenic pathogen in the environment, allowing it to colonize many new individuals.

Mucosal surfaces are usually the first host surface pathogens contact. The ability of pathogens to successfully adhere to cellular surfaces is a key virulence attribute [12]. Adherence is usually mediated by various bacterial adhesins such as fimbriae, pili, and afimbrial adhesins. Epithelial cells of the intestinal surface (and most other epithelial surfaces) form extensive networks of actin-based microvilli. Although this increases the cellular surface area, which enhances nutrient uptake, it is unclear whether this is a benefit or detriment to bacterial adherence. Nonetheless, some bacterial pathogens trigger microvillus disruption and surface rearrangements, including *Salmonella* species and the attaching and effacing pathogens such as pathogenic *E. coli* [6]. Because actin is the key regulator of microvilli, it is not surprising that these pathogens target small Rho GTPases [8] to disrupt microvilli. In addition, *Salmonella* is able to insert a phosphatidyl inositate phosphate phosphatase (SigD) into the host cell, which "loosens" the cellular surface (presumably by altering the underlying cytoskeletal architecture), enhancing bacterial invasion [17]. Given their key role in cellular signaling, the ability to modulate a host cellular surface could be of significant benefit to a pathogen intent on adhering to an epithelial surface. This could include up- or downregulating one or more host cell proteins with which the pathogen might interact (see chapter by Duménil and Nassif, this volume). It could also include altering cellular surface function that may impact on the host's response to a pathogen (such as affecting Toll-like receptor signaling normally used to detect pathogens). Finally, it could also include altering the underlying cytoskeletal architecture. For example, Rho GTPases modulate the ezrin/radixin/moesin proteins in leukocytes, which affects integrin, ICAM, L-selectin, and other surface markers, which ultimately affects leukocyte migration, T-cell interaction with antigen-presenting cells, apoptosis, and phagocytosis [13].

Perhaps one of the most remarkable attributes of pathogens that activate small G proteins is that they often also encode the machinery to turn the G proteins off once the process is complete. This has been worked out very well for *Salmonella* species (see chapter by Schlumberger and Hardt, this volume, and [8]). This pathogen uses a type III secretion system to inject at least three bacterial effectors [SopE (a GEF), SopE2 (also a GEF), and SigD], which activate Cdc42 and Rac to facilitate actin-based bacterial uptake and invasion. However, it uses the same secretion system to also insert SptP, a GAP that inactivates Cdc42 and Rac activity and returns the actin cytoskeleton to a near-normal state after bacterial invasion, while the bacteria remain inside a vacuole in the host cell.

The ability to enter into a host cell provides a unique opportunity for a pathogen. It allows it to enter a noncompetitive environment free from normal flora. (In the large bowel, there is such competition for nutrients that *E. coli*'s division time is about 24 h, compared to 20 min in rich broth, and it is in stationary phase in the gut). In addition, once inside a host cell, a pathogen is privy to a rich nutrient and moisture source, free from circulating antibodies and patrolling macrophages and neutrophils. Many bacterial pathogens invade host cells, and nearly all do this by exploiting members of the Rho GTPase family [5]. Most, such as *Shigella* and *Salmonella*, do this by injecting effectors that modulate Rac and Cdc42 activity. Examples of pathogens that use Rho family-mediated invasion include *Bartonella, Brucella, Chlamydia, Listeria,* and *Pseudomonas*. Indeed, it is felt that nearly all invasive pathogens use such a strategy for invasion, except perhaps for *Rickettsia*, which seems to use a phospholipase activity to break into a cell.

Once inside a cell, some pathogens digest the vacuole surrounding them and escape into the cytoplasm, where they then polymerize cytosolic actin [4]. This polymerization event propels them around inside the host cell and enables them to spread into adjacent host cells without becoming extracellular. *Shigella, Listeria,* and some *Rickettsia* use such a mechanism as part of their virulence [4]. However, instead of activating small Rho GTPases, these pathogens actually mimic downstream components to achieve this actin rearrangement. The *Shigella* protein IcsA recruits and activates N-WASP, leading to Arp2/3 complex recruitment and actin polymerization. This thus replaces the function of Cdc42. The *Listeria* protein ActA activates Arp2/3 directly, thereby directly mimicking N-WASP's activity.

Another family of pathogens, including enteropathogenic *E. coli* (EPEC) and enterohemorrhagic *E. coli* (EHEC), uses a type III system to insert effectors and their own receptors into epithelial surfaces. This results in a spectacular rearrangement of cellular actin and pedestal formation beneath the pathogen. Although Rho G proteins do not seem to be involved in this process [2], EPEC inserts its own protein (Tir) into host membranes, where it is tyrosine phosphorylated and binds the adaptor protein Nck, which then activates N-WASP and Arp2/3 complex recruitment [9]. The above examples indicate that modification or activation of Rho G proteins is not necessary, but instead mimics can be deployed to modulate actin dynamics.

For pathogens that invade and remain within a vacuole, it is critical to avoid lysosomal fusion (and subsequent bacterial death). Again, Rho GTPases (as well as Rabs) are involved in this process and other intracellular killing mechanisms, and inactivation of these may enhance intracellular survival (see chapter by Diebold and Bokoch, this volume). For example, *Legionella pneumophila* uses a type IV secretion system to insert an effector (RalF) into host cells that functions as a GEF to activate the small GTPase ARF, which modulates lysosomal fusion of the vacuoles containing bacteria [15].

Professional phagocytic cells are designed to use actin-mediated events to internalize pathogens and then destroy them in phagolysosomes. However, several pathogens including *Yersinia*, *Pseudomonas*, and EPEC have developed strategies to avoid phagocytosis by blocking this process [3]. Given the central role of actin in phagocytosis, it is no surprise that these mechanisms involve pathways to disrupt normal actin-mediated processes (see chapter by Deckert et al., this volume).

Many bacterial toxins and effectors either activate or deactivate Rho GTP-ases for a variety of functions [1]. Often this activity is mediated by enzymatic modification of the G protein. The variety of effects modulated by these bacterial toxins is impressive and serves as a major arsenal for pathogenic microbes. Often killing the host cell is one outcome, but many more subtle effects have been noted, which are a significant focus of this volume. In addition, these effectors and toxins have become valuable tools to dissect normal cellular functions and are used extensively by cell biologists without realizing their natural role in biology.

4
Looking Ahead

Although many examples of virulence strategies that involve Rho GTPases have been documented, there are many, many more undiscovered ones. In fact, it is probable that for every normal and important cellular process, there is probably a pathogen that has designed a mechanism to exploit it. There is currently an explosion in identifying type III and type IV effectors that are inserted into host cells yet currently lack an identified host target. It is probable that, in addition to the Rho family, there are many intracellular pathogens that alter members of the Rab and Arf family to survive inside host cells; this is a field in its infancy. The extensive signaling mediated by the Ras family makes these ideal targets for virulence factors to reprogram cells. It is also likely that factors that alter Ran function and reprogram the nucleus will be identified.

It is apparent that there are several examples of bacterial virulence factors that specifically target members of the Rho GTPase family and their regulators. What is remarkable is the diverse effects such mechanisms have on virulence strategies, ranging from killing the host cell to mediating invasion and intracellular invasion. Because of the genetic plasticity and promiscuity of bacterial genetics, once a pathogen has found a successful virulence mechanism, this is rapidly passed on by plasmids, phage, and conjugation to other pathogens. This results in new combinations of virulence factors, which subsequently translate into new pathogens and emerging infectious diseases. Given the extent of bacterial modulators of Rho GTPases, there is no doubt that many more such effectors will be discovered. Also, because these GTPases all use similar mechanisms for activation and regulation (Fig. 1), it is not hard to imagine how a pathogen might alter a virulence factor to work in a similar way on a different GTPase.

The ability to cause disease is a compilation of mechanisms that collectively reprogram and override the host, resulting in pathogen proliferation and spread. Because Rho GTPases play such a central role in normal cellular function, they are used by many successful bacterial pathogens as targets of effectors. As our knowledge increases about these mechanisms, so does our understanding of bacterial virulence. An added benefit is that these toxins and effectors also serve as excellent tools to understand basic cellular processes. This volume is a testimony to this rapidly expanding and important knowledge.

Acknowledgements B.B.F. is a Howard Hughes Medical Institute (HHMI) International Research Scholar, a Canadian Institutes for Health Research (CIHR) Distinguished Investigator, and the UBC Peter Wall Distinguished Professor. Operating grants from HHMI, CIHR, and the Canadian Bacterial Diseases Network support work in the author's laboratory.

References

1. Aktories, K., G. Schmidt, and I. Just. 2000. Rho GTPases as targets of bacterial protein toxins. Biol Chem 381:421–6.
2. Ben-Ami, G., V. Ozeri, E. Hanski, F. Hofmann, K. Aktories, K. M. Hahn, G. M. Bokoch, and I. Rosenshine. 1998. Agents that inhibit Rho, Rac, and Cdc42 do not block formation of actin pedestals in HeLa cells infected with enteropathogenic *Escherichia coli*. Infection and Immunity 66:1755–8.
3. Celli, J., and B. B. Finlay. 2002. Bacterial avoidance of phagocytosis. Trends Microbiol 10:232–7.
4. Cossart, P. 2000. Actin-based motility of pathogens: the Arp2/3 complex is a central player. Cell Microbiol 2:195–205.
5. Cossart, P., and P. J. Sansonetti. 2004. Bacterial invasion: the paradigms of enteroinvasive pathogens. Science 304:242–8.
6. Finlay, B. B., and P. Cossart. 1997. Exploitation of mammalian host cell functions by bacterial pathogens. Science 276:718–725.
7. Finlay, B. B., and S. Falkow. 1997. Common themes in microbial pathogenicity revisited. Microbiol Mol Biol Rev 61:136–69.
8. Galan, J. E., and D. Zhou. 2000. Striking a balance: modulation of the actin cytoskeleton by *Salmonella*. Proc Natl Acad Sci USA 97:8754–61.
9. Gruenheid, S., R. DeVinney, F. Bladt, D. Goosney, S. Gelkop, G. D. Gish, T. Pawson, and B. B. Finlay. 2001. Enteropathogenic *E. coli* Tir binds Nck to initiate actin pedestal formation in host cells. Nat Cell Biol 3:856–9.
10. Hall, A. 1998. Rho GTPases and the actin cytoskeleton. Science 279:509–14.

11. Hueck, C. J. 1998. Type III protein secretion systems in bacterial pathogens of animals and plants. Microbiol Mol Biol Rev 62:379–433.
12. Hultgren, S. J., S. Abraham, M. Caparon, P. Falk, J. St Geme, and S. Normark. 1993. Pilus and nonpilus bacterial adhesins: assembly and function in cell recognition. Cell 73:887–901.
13. Ivetic, A., and A. J. Ridley. 2004. Ezrin/radixin/moesin proteins and Rho GTPase signalling in leucocytes. Immunology 112:165–76.
14. Koster, M., W. Bitter, and J. Tommassen. 2000. Protein secretion mechanisms in Gram-negative bacteria. Int J Med Microbiol 290:325–31.
15. Nagai, H., J. C. Kagan, X. Zhu, R. A. Kahn, and C. R. Roy. 2002. A bacterial guanine nucleotide exchange factor activates ARF on *Legionella* phagosomes. Science 295:679–82.
16. Takai, Y., T. Sasaki, and T. Matozaki. 2001. Small GTP-binding proteins. Physiol Rev 81:153–208.
17. Terebiznik, M. R., O. V. Vieira, S. L. Marcus, A. Slade, C. M. Yip, W. S. Trimble, T. Meyer, B. B. Finlay, and S. Grinstein. 2002. Elimination of host cell PtdIns(4,5)P(2) by bacterial SigD promotes membrane fission during invasion by *Salmonella*. Nat Cell Biol 4:766–73.

CTMI (2005) 291:11–28

Extracellular Bacterial Pathogens and Small GTPases of the Rho Family: An Unexpected Combination

G. Duménil (✉) · X. Nassif

INSERM Unité 570, Faculté de Médecine Necker-Enfants Malades, 156 rue de Vaugirard, 75015 Paris, France
dumenil@necker.fr

Abstract Even in the case of extracellular bacterial pathogens, it is becoming increasingly clear that successful colonization does not limit itself to passive attachment on the surface of human cells; a dialogue takes place between bacteria and infected cells.

These pathogens modulate cellular functions to their advantage, leading to survival and proliferation at the cell surface. Furthermore, there is increasing evidence that a variety of extracellular pathogens activate small GTPases of the Rho family during adhesion, placing these regulators at the center of the interaction between these bacteria and their infected host.

1 Introduction

Small GTPases of the Rho family are central regulators of actin cytoskeleton dynamics. Their levels of activation coordinate actin cytoskeleton organization, leading to the formation of distinct actin-based structures that guide changes in cellular morphology (Etienne-Manneville and Hall 2002). As such, they are in the center of numerous cellular functions such as cellular motility (Raftopoulou and Hall 2004), phagocytosis (Chimini and Chavrier 2000), cell polarity (Van Aelst and Symons 2002), or intracellular trafficking (Symons and Rusk 2003). Bacterial pathogens often target small GTPases of the Rho family to divert cellular functions to their advantage (see other chapters in this volume). For pathogens with an intracellular lifestyle, such as *Salmonella*

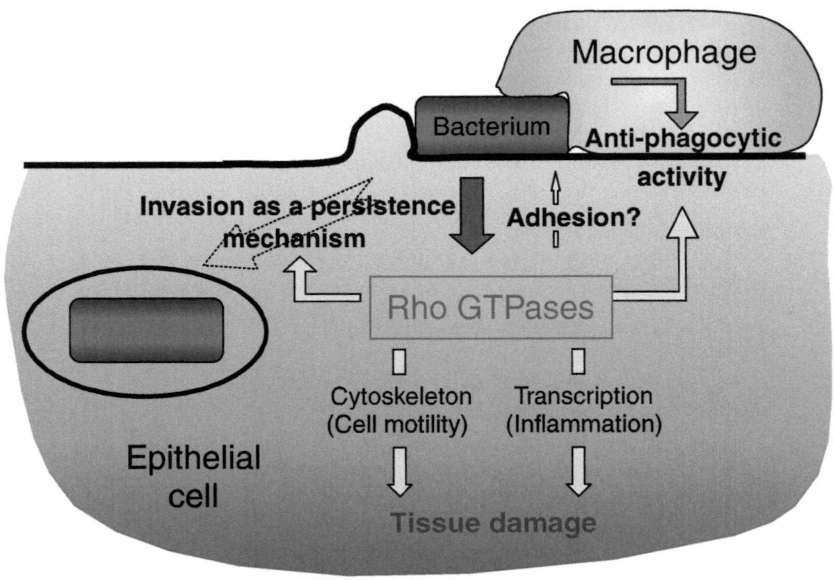

Fig. 1 Rho family GTPase activation by extracellular bacteria and possible consequences

typhimurium, *Shigella flexneri*, or *Listeria monocytogenes*, the activation of small GTPases is a prerequisite to bacterial internalization, a process that allows optimal growth conditions and protects bacteria from the immune system. For extracellular bacteria, however, even though there is increasing evidence that small GTPases are activated during infection, the significance of small GTPase activation remains unclear. The goal of this review is to describe what is known of the mechanisms of activation of Rho family proteins by extracellular bacteria and to discuss the biological significance of this activation. Pathogens such as *Neisseria meningitidis*, *Pseudomonas aeruginosa*, uropathogenic *E. coli* (UPEC), enteropathogenic *E. coli* (EPEC), and *Helicobacter pylori* were chosen as paradigms of extracellular bacterial pathogens. Different aspects of the interaction between these bacteria and host cells will be discussed: adhesion in the literal sense of bacteria sticking to cells, bacteria-induced epithelium damage, low-level host cell invasion, and resistance to phagocytosis (see Fig. 1).

2 Extracellular Pathogens Are Potent Activators of Small GTPases of the Rho Family

The activation of small GTPases during the interaction of extracellular pathogens with host cells has, in some cases, been directly demonstrated with the binding specificity of CRIB (Cdc42/Rac interactive binding) domains toward the GTP-bound form of Rho family proteins (Ren et al. 1999). Alternatively, indirect evidence through the use of inhibitors of small GTPases such as dominant negative forms or chemical inhibitors indicates that they are involved in a given process of infection.

2.1 *P. aeruginosa* Activates Rho or Cdc42 Depending on the State of Differentiation of Epithelial Cells

The normal nonspecific defenses of the human body are sufficient to prevent *P. aeruginosa* from causing infections, but serious breaches of these defenses (burns, immunosuppressive therapy, or wounds) can allow *P. aeruginosa* to invade the body and cause fatal systemic disease. The expression of a type IV pilus at the surface of the bacteria allows *P. aeruginosa* to adhere to epithelial cells at the port of entry. A type III secretion system then allows direct delivery of effector proteins into the host cell, thus causing further damage. In vitro,

P. aeruginosa infection of partially differentiated MDCK cells increases the level of GTP-bound (activated) RhoA with a peak at 60 min (Kazmierczak et al. 2004). On the other hand, the same *P. aeruginosa* strain activates Cdc42 but not Rho when MDCK cells are differentiated, with an activation peak at 20 min. The molecular basis for this difference is not clear but has interesting implications as *P. aeruginosa* specifically targets damaged (undifferentiated) epithelia.

2.2
H. pylori, Type IV Secretion-Dependent Activation of Rac and Cdc42

H. pylori is associated with chronic superficial gastritis (stomach inflammation) and plays a role in the pathogenesis of peptic ulcer disease. Increasing evidence also indicates that *H. pylori* infection is important in promoting gastric carcinoma and lymphoma. Most of the time, however, chronic infection is asymptomatic. Despite gastric acidity, *H. pylori* proliferates in the mucus layer and a small proportion of cells (10%) adheres to the gastric epithelium. The microorganism does not appear to invade tissue. Production of both a vacuolating cytotoxin (VacA) and the *cag*-encoded type IV secretion system is associated with injury to the gastric epithelium. In vitro, *H. pylori* infection activates Rac1 and Cdc42 1 h after infection (Churin et al. 2001). The *cag*-encoded secretion system is necessary for this process, suggesting that bacterial effectors secreted by the *cag* secretion system are responsible for the activation. CagA is the only known substrate of this secretion system, but this protein is not involved in this process, because a *cagA* mutant triggers the activation of small GTPases. Other secreted factors are therefore involved in activating the small GTPases on *H.pylori* infection. Alternatively, it has been proposed that the secretion system itself could be responsible for activation of the Rho GTPases by interacting with a surface receptor (Naumann et al. 1999).

2.3
N. meningitidis-Induced Cytoskeletal Rearrangements Are Cdc42- and Rho Dependent

N. meningitidis is responsible for septicemia and/or meningitis. Infection occurs after aspiration of infective particles and colonization of the nasopharynx. Disease is a consequence of bloodstream invasion and the crossing of the blood–brain barrier. Bacterial adhesion is thought to be the initial step leading to internalization and transcytosis of the bacteria through the endothelial cells of the brain (Nassif et al. 2002). *N. meningitidis* express type

IV pili at their surface that allow efficient bacterial adhesion and the formation of microcolonies on the surface of host cells. Depending on the isolate, *N. meningitidis* also express different adhesins of the Opa or Opc family. In vitro, *N. meningitidis* trigger a potent cytoskeletal reorganization under the microcolonies and exhibit a low level of invasion. Bacterial invasion and the formation of the cytoskeleton modifications can be inhibited by expressing dominant negative forms of Rho and Cdc42 but not Rac (Eugene et al. 2002).

2.4
Enteropathogenic *E. coli* and Enterohemorrhagic *E. coli* Induce Pedestal Formation Independently of Rho Family GTPases

Diarrhea caused by the enteroinvasive, cytotoxic, and enteropathogenic (EPEC) strains of *E. coli* ranges from very mild to severe. EPEC also possess a type IV pilus (termed bundle-forming pilus, or BFP). BFP are responsible for the initial attachment of EPEC to intestinal target cells. Attachment is followed by the destruction of microvilli, a process known as attaching and effacing (A/E). This initial phase of attachment is followed by the formation of pedestals supporting individual bacteria. In the case of EPEC and enterohemorrhagic *E. coli* (EHEC), pedestal formation is independent of small GTPases but relies nevertheless on the induction of actin reorganization pathways involving N-WASP and Arp2/3. In contrast, Cdc42 inhibition enhances pedestal formation (Ben-Ami et al. 1998; Kenny et al. 2002). EPEC were also shown to transiently induce other types of cellular projections that are linked to the invasive property of the bacteria (Kenny et al. 2002). Both invasion and cellular projections can be inhibited by dominant negative Cdc42, pointing to a role of this small GTPase in this process.

2.5
UPEC Activates Rho, Rac, and Cdc42

Uropathogenic *E. coli* is the primary cause of urinary tract infections. Infections caused by UPEC are usually self-limiting and rarely spread beyond the urinary tract but are often recurrent. The most important virulence factor for these bacteria is the enhanced ability to adhere to uroepithelial cells (Sauer et al. 2000). This attachment is mediated by a type I pilus on the surface of *E. coli*; the specific pilus adhesins located at the pilus tip mediate interaction with cellular oligosaccharides. A small proportion of adhering UPEC are internalized into epithelial cells of the urinary tract (Martinez et al. 2000). This invasion process is dependent on Rho, Rac, and Cdc42 (Martinez and Hultgren 2002).

3 Small GTPase Activation and Bacterial Adhesion in the Literal Sense

EPEC or EHEC reorganize the underlying cytoskeleton to generate a short and wide cellular projection that is intimately associated with bacteria (Campellone and Leong 2003). The biological significance of pedestal formation remains unclear, although it has been suggested that these pedestals could mediate "firm adhesion" (Vallance and Finlay 2000). The anchoring of the bacteria to the host cytoskeleton could strengthen the adhesion; however, this remains to be demonstrated. In N-WASP-deficient cells, for example, pedestals do not form but EPEC readily bind to cells (Lommel et al. 2001). During meningococcal interaction with host cells, inhibition of small GTPase of the Rho family by ToxB did not lead to any quantifiable change in pili-dependent bacterial adhesion (Eugene et al. 2002; Kallstrom et al. 2000). Compactin, a general small GTPase inhibitor, inhibited UPEC invasion but did not affect UPEC adhesion capacity (Ben-Ami et al. 1998). In summary, there is no clear evidence for a role of the actin cytoskeleton in modifying the number of bacteria adhering to host cells. It is possible, however, that conditions used in the laboratory are not stringent enough to test the role of the cytoskeleton in the adhesion process. Under physical stress such as hydrodynamic fluxes found in vivo, cytoskeleton reorganization could play a key role in strengthening the interaction (Thomas et al. 2002).

4 Small GTPase Activation and Epithelium Destruction

It is not clear how tissue damage affects the infection process, but it could be important for evasion of the immune system or access to deeper tissues or for favoring bacterial dissemination to another host. In any case, it is thought to be involved in *H. pylori*-triggered pathological symptoms associated with epithelium destruction such as peptic ulcer and could be involved in cancer development, as the morphological changes induced by *H.pylori* resemble the process of oncogenic transformation.

4.1 *H. pylori* Infection, Inflammation, and Epithelial Cell Scattering

An important part of tissue damage triggered by *H. pylori* is thought to occur as an indirect consequence of the inflammation induced by the bacteria. In addition, *H. pylori* plays a more direct role in this process by triggering the dissociation, migration, and remodeling of epithelial monolayers, A process reminiscent of the effects induced by the hepatocyte growth factor (HGF) (Churin et al. 2001; Segal et al. 1999).

4.1.1 *H. pylori*-Triggered Signaling Leading to Gene Transcription Regulation and Inflammation

An important aspect of small GTPase biology is the initiation of signaling cascades leading to gene regulation. The activation of the GTPases of the Rho family leads to the activation of NF-κB and AP-1 transcription factors, which are central to the transcriptional regulation of numerous genes including those involved in the control of inflammation. This pathway is believed to play an important role in the inflammation induced by *H. pylori* adhesion to epithelial cells.

The immune response to *H. pylori* infection is initiated by a number of inflammatory mediators including cytokines and chemokines produced from the gastric epithelium. In vivo and in vitro studies have shown that *H. pylori* induces the production of chemokines, IL-8, RANTES, GRO-α, MIP1-α, ENA-78, and MCP-1 and cytokines IL-1, IL-6, and TNFα (Bodger and Crabtree 1998). Inflammatory mediators produced by polymorphonuclear leukocytes and mononuclear phagocytes recruited to the site of infection could directly damage the surface epithelial layer, leading to loss of microvilli, irregularity of the brush border, and vacuolation. Consistently, the inflammation process requires de novo synthesis of chemokines and proinflammatory cytokines. *H. pylori* infection leads to activation of the transcription factors AP-1 and NF-κB after 90 min of infection (Naumann et al. 1999). The *cag* pathogenicity island was shown to be an important factor triggering this activation (Glocker et al. 1998). The *H. pylori*-triggered signaling pathway leading to AP-1 and NF-κB activation is composed of JNK, MKK4, PAK, and Rho GTPases (Naumann et al. 1999). Expression of dominant negative forms of Cdc42 and Rac leads to a sharp decrease in AP-1-dependent gene activation after infection.

Global approaches using DNA microarray technology revealed that the transcription of numerous genes is modulated by *H. pylori* infection (Guillemin et al. 2002). Among the genes upregulated by *H. pylori* is the matrix metalloprotease MMP-7 (matrilysin) (Wroblewski et al. 2003). Levels of MMP-7 are high in biopsies from patients infected with *H. pylori*. MMP-7 is commonly upregulated in gastric cancer, thus suggesting that *H. pylori*-induced levels of MMP-7 could be an important factor in bacteria-induced neoplasia (Crawford et al. 2002; McDonnell et al. 1991). Consistent with this, antisense to MMP-7 inhibited bacteria-induced cell spreading. Induction of MMP-7 does not seem to be through a paracrine effect but rather a direct effect, and dominant negative forms of Rac and Rho inhibited *H. pylori* induction of the MMP-7 gene at the transcriptional level (Wroblewski et al. 2003).

4.1.2
H. pylori and the Induction of Epithelial Cell Scattering

After *H. pylori* infection, epithelial cells spread and detach 1 h after infection, acquiring a typical cellular morphology known as the hummingbird. The mechanism by which *H. pylori* trigger epithelial cells scattering has recently been the focus of several studies. An important observation is that *H. pylori*-dependent scattering is not dependent on de novo protein synthesis (cycloheximide resistant) whereas HGF-triggered scattering requires protein synthesis (cycloheximide sensitive) (Selbach et al. 2003). In contrast to HGF, *H. pylori* could therefore exert its effect by directly targeting regulators of the cytoskeleton such as Rho family GTPases. Consistent with this, mutants in the *cag* secretion system fail to induce cell motility correlating with the absence of small GTPase activation, but a direct link between both processes remains elusive. It is clear, however, that *H. pylori*-induced small GTPase activation is not sufficient to trigger cell motility. *cagA* mutants still activate GTPases but fail to trigger efficient motility (Churin et al. 2003), thus suggesting that CagA is an important player in this process. CagA seems to interfere with cellular signaling at several levels that could account for its role in cell motility. CagA was found to bind to and activate the c-met/HGF receptor (Churin et al. 2003) and could stimulate cellular motility through the modulation of HGF receptor activity. CagA function inside cells is also tightly linked to the function Src tyrosine kinase; CagA is phosphorylated by this kinase and, once phosphorylated, inhibits Src tyrosine kinase activity, leading to decreased phosphorylation of cortactin (Selbach et al. 2003). CagA was also shown to interact with the tyrosine kinase CSK (Tsutsumi et al. 2003). Alteration of tyrosine kinase-dependent signaling pathways could thus be another mode of action of CagA. In summary, it can be hypothesized that CagA acts in conjunction with other secreted factors that stimulate the small GTPases of the Rho family to induce cell scattering.

4.2
P. aeruginosa Inhibition of Wound Healing

Epithelium injury and the undifferentiated state of injured epithelial cells favor *P. aeruginosa* adhesion and invasion (Fleiszig et al. 1997). In turn, the consequence of this adhesion process is the inhibition by *P. aeruginosa* of epithelial wound repair. This inhibition is primarily localized at the edge of the wound and is linked to the collapse of actin cytoskeleton, i.e., cell rounding and cell detachment. The *P. aeruginosa* effector for this process is ExoT, which is secreted by the type III secretion system. ExoT is highly homologous to ExoS, the other effector of the *P.aeruginosa* type III secretion system (74% identity at the amino acid level). Both are bifunctional proteins (Yahr et al. 1996); one domain has an ADP-ribosylation activity, and the other stimulates GTPase activity of Rho GTPases (GAP activity). ExoT-dependent inhibition of wound repair is mediated through the GAP activity of this bacterial protein, as mutations in ExoT that alter the conserved arginine (R149) within the GAP domain abolish the ability of *P. aeruginosa* to inhibit wound closure (Geiser et al. 2001).

4.3
EPEC, Malabsorption, and Diarrhea

The lesion caused by EPEC consists mainly of destruction of microvilli during the attaching and effacing lesion. Loss of microvilli leads to malabsorption and osmotic diarrhea. This process results from the secretion of protein effectors through the type III secretion system that target the actin cytoskeleton but not Rho family GTPases.

5
Inhibition of Phagocytosis

To persist, extracellular bacterial pathogens have to circumvent phagocytosis by polymorphonuclear leukocytes and macrophages. Phagocytosis involves the formation of cellular projections that engulf the bacteria, and this step requires activation of Rho family members (Chimini and Chavrier 2000). Depending on whether the exogenous particles to be phagocytosed are opsonized by complement or by antibodies, different signaling pathways are triggered. Complement-dependent phagocytosis requires the small GTPase Rho, whereas Fc receptor mediated phagocytosis is mediated by Cdc42 and Rac in addition to tyrosine kinases (Caron and Hall 1998). Bacteria with antiphagocytic activity once again target small GTPases of the Rho family (Ernst 2000). The antiphagocytic effect of *Yersinia spp.* is well characterized and is the object of a chapter in this volume and will not be developed here. Other extracellular bacteria have antiphagocytic activity; among them are *P. aeruginosa* and *H. pylori.*

5.1
P. aeruginosa ExoS and ExoT as Inhibitors of Phagocytosis

As for *Yersinia spp.*, *P. aeruginosa* inhibits phagocytosis through the secretion of protein effectors (ExoS and ExoT in the case of *P. aeruginosa*) into epithelial cells and macrophages by the type III secretion system. Both ADP-ribosylating and GAP activities of ExoS and ExoT can account for antiphagocytic properties. The GAP activity of ExoS and ExoT is effective on Rho GTPase family members and can by itself inhibit phagocytosis (Kazmierczak and Engel 2002; Krall et al. 2000). In addition, ExoS ADP-ribosylates numerous proteins including members of the ras, rab, and ral families of GTPases, which accounts for its cytotoxic activity as well as antiphagocytic activity (Barbieri et al. 2001; Ganesan et al. 1998). ExoT ADP-ribosylating activity exhibits a different specificity and targets Crk family members (Sun and Barbieri 2003). Crk proteins are adaptors containing SH2 and SH3 domains involved in the signaling regulating several cellular processes including phagocytosis (Matsuda et al. 1992).

5.2
H. pylori, Partial Inhibition of Phagocytosis and Intracellular Survival

An increased number of T lymphocytes, macrophages, and polymorphonuclear leukocytes are evident in histological sections of gastric mucosa from patients with *H. pylori* infections (Kazi et al. 1989). Despite the presence of these phagocytes, *H. pylori* can persist several decades, suggesting the existence of a resistance mechanism to phagocytosis. *H.pylori* is capable of inhibiting its internalization into the host cells (Ramarao et al. 2000). This requires the presence of the type IV secretion system. The effector responsible for this effect is not known, as *cagA* mutants maintain the ability to inhibit phagocytosis. In addition to their antiphagocytic activity, internalized *H. pylori* are able to survive and persist inside peritoneal macrophages (Allen et al. 2000). The *cag* secretion system is also necessary for intracellular survival, as type II strains (which do not carry the *cag* pathogenicity island) are killed by macrophages. In this case also, the type IV secretion system effector and its cellular target are not known. Interestingly, as mentioned above, it has been shown that *H. pylori* activates Cdc42 and Rac, suggesting that these two small GTPases are not the target of the antiphagocytic effect of *H. pylori*.

6
Bacterial Invasion as a Persistence Mechanism

Although UPEC, EPEC, and *N. meningitidis* are described as bacteria with predominantly extracellular multiplication, a relatively small proportion of bacteria can be found in intracellular locations both in vivo and in vitro. The significance of these observations generally remains unclear except in the case of UPEC, where elegant studies in a mouse bladder infection model demonstrated that intracellular bacteria are key for persistence and reinfection.

6.1
UPEC Persistence in a Mouse Cystitis Model

In vitro experiments first suggested that UPEC may be intracellular, and gentamicin invasion assays revealed that about 3% of cell-associated UPEC are intracellular after 1 h of infection (Martinez et al. 2000). Invasion requires the expression of the type I pilus and, more specifically, the FimH adhesin at the tip of the pilus. Furthermore, this protein coated on beads is sufficient for efficient internalization (50% internalized beads). Internalization is inhibited by compactin, a general inhibitor of Rho GTPases. Rho, Rac, and Cdc42 are all necessary for bacterial internalization, as C3 toxin, N17Cdc42 and N17Rac1 each decrease internalization three- to fourfold (Martinez and Hultgren 2002). Similar results were observed with FimH-coated beads. With the use of constitutively active small GTPases, it has also been demonstrated that tyrosine kinases and PI3-kinases are likely to act upstream of Rac and downstream of Cdc42 in the pathway leading to internalization of UPEC.

The low level of internalization of the bacteria in cell culture models raises the issue of the relevance of this phenomenon. This issue has been addressed with a mouse cystitis model (Mulvey et al. 1998, 2001). The luminal surface of the bladder is lined by a layer of superficial umbrella cells that deposit on their apical surface a quasi-crystalline array of hexagonal complexes made of four integral membrane glycoproteins called uroplakins. Type I pili were shown to be necessary for bladder colonization as they can bind to uroplakins on the surface of the umbrella cells. Interestingly, bacteria are often found in membrane ruffles, suggesting that bacteria stimulate membrane dynamics in these conditions.

A kinetic study revealed that total bacterial counts in the bladder rapidly decreased after inoculation by transurethral catheterization, presumably due to exfoliation of the epithelium. Furthermore, cellular exfoliation is dependent on an apoptotic process. The proportion of intracellular bacteria was determined with a gentamicin assay on whole bladders dissected from the infected animals. Two hours after inoculation, the proportion of intracellular bacteria is low (0.2%) but after 12 h, most remaining bacteria are intracellular and nearly all of them are intracellular at 48 h. These studies clearly demonstrate that the low amount of internalization leads to persistent infection and are consistent with the recurrent characteristic of cystitis. It is interesting that intracellular bacteria are not cleared by exfoliation, suggesting that this process is solely induced by extracellular bacteria and that, once inside, UPEC are not sensed by the cells, thus revealing an unexpected advantage to bacterial internalization.

6.2
N. meningitidis, Persistence in the Nasopharynx

In the case of *Neisseria meningitidis*, bacterial adhesion is mediated by type IV pili expressed at the surface of the bacteria. CD46 (membrane cofactor protein) is proposed as the cellular receptor for bacterial pili (Kallstrom et al. 1997). Adhesion leads to powerful actin cytoskeleton reorganization under the bacterial colony (Merz et al. 1999). In addition to actin itself, cytoskeletal proteins such as ezrin or cortactin are recruited at the site of colony attachment. Several transmembrane receptors are clustered under the colony, including CD44, ICAM-I, or ErbB2 (Eugene et al. 2002; Hoffmann et al. 2001). It is thought that the intense actin cytoskeleton reorganization guides the cellular projections observed by electron microscopy under and around bacterial colonies. It is noteworthy that these cellular projections have been observed in organ cultures on binding to epithelial cells and on endothelial cells in histopathological studies (Nassif et al. 2002; Stephens et al. 1983). It has been proposed that these projections are a step toward the observed low-efficiency bacterial internalization (0.1% of adherent bacteria) (Eugene et al. 2002). Inhibition of Rho or Cdc42 leads to a decrease in the ability of the bacteria to reorganize the actin cytoskeleton under the colony, thus correlating with lower numbers of intracellular bacteria (Eugene et al. 2002). Surprisingly, dominant negative forms of Rac failed to affect either actin reorganization or internalization, suggesting that Rho and Cdc42 are the main regulators of this process.

N. meningitidis is strictly a human pathogen, and the absence of good animal models has considerably impeded our understanding of the pathogenesis process caused by this bacteria. A few studies have used organ culture models with biopsy samples (Rayner et al. 1995) but most are based on interaction between bacteria and cells in culture. An interesting study used clinical samples obtained from tonsillectomies (Sim et al. 2000). Samples were processed for immunohistochemistry and labeled with antibodies directed against neisserial epitopes. The main conclusion of the study was that asymptomatic carriage might be largely underestimated by studies using only nasopharyngeal swabbing. Immunohistochemistry detected *N. meningitidis* in nearly half of the samples, whereas nasopharyngeal swabbing only in 10% of patients. The authors made an interesting observation that in some cases (4/32) bacteria could be detected within or beneath the epithelium surface. This conclusion was based on colabeling with cytokeratin and confocal observation. These observations suggest that, as in the case of UPEC, *N. meningitidis* could use an intracellular niche to maintain persistent colonization of the epithelium. The low level of internalization observed in vitro could thus have high significance in vivo. In rare instances, this survival strategy could be harmful to the survival of the bacterium through the development of invasive disease and death of the host.

6.3 EPEC, Filopodia Formation, and Invasion

Several publications reported in vitro and sometimes in vivo internalization of EPEC or EHEC strains (Donnenberg et al. 1989; Donnenberg and Kaper 1992). Conditions can be found in which reproducible invasion of about 10% of adhering bacteria occurs after 60-min infection (Jepson et al. 2003). Although the clinical significance of this process is not clear, most isolates present this property (Donnenberg et al. 1989) and several studies have attempted to identify the mechanisms involved in this process. It was first noticed that pedestals are only one of the cellular structures induced by EPEC. In particular, several studies point to the existence of short-lived filopodia at the site of attachment of EPEC colonies (Kenny et al. 2002; Phillips et al. 2000). Filopodia are formed 5 min after attachment and disappear 15–20 min before pedestal formation (Kenny et al. 2002). These filopodia are proposed to play an important role in EPEC invasion.

On pilus-mediated adhesion, EPEC secrete various proteins including the proteins Tir and MAP (mitochondrion-associated protein) through a type III secretion system. The MAP protein seems to be a central player in EPEC-induced filopodia formation (Kenny et al. 2002; Kenny and Jepson 2000). The MAP protein is also involved in EPEC internalization, as MAP mutants are less invasive and MAP overexpression leads to a hyperinvasive phenotype. Tir, on the other hand, inserts into host cells and serves as a receptor for the intimin protein at the surface of the bacteria (Kenny et al. 1997), and this interaction triggers pedestal formation (Rosenshine et al. 1996). Tir/intimin interaction is also necessary for internalization, although it is not necessary for filopodia formation (Jepson et al. 2003). Tir is thus involved in both pedestal formation and invasion. Interestingly, the two properties of Tir can be dissociated: A Y474S mutant does not form pedestals but still induces filopodia, and an R521A mutation does not form filopodia but still forms pedestals, suggesting that two different signaling pathways are involved (Kenny et al. 2002). On the cellular side, Cdc42 but not Rac is necessary for EPEC internalization.

7 Conclusion

At first glance, it is not obvious how bacteria adhering to an epithelium should benefit from modulation of the host cell cytoskeleton, and there is no evidence that small GTPases of the Rho family influence bacterial adhesion in the literal sense. However, if one considers adhesion in a broader sense, including survival and proliferation at the surface of an epithelium, the role of small GTPases of the Rho family becomes apparent. Bacteria such as *P. aeruginosa* and *H. pylori* trigger tissue damage on tissue colonization, suggesting that tissue disorganization might favor their ability to colonize or to disseminate to another host. It appears also that there is a delicate balance between invasion and antiphagocytic activity. Certain bacteria such as *N. meningitis* or UPEC express factors that favor a low level of invasion that could influence long-term persistence. In contrast, other bacteria inhibit phagocytosis probably as a mechanism of evasion from phagocytic cells. These properties are not mutually exclusive, and it is noteworthy that *Yersinia spp.* express both an invasin protein that triggers efficient internalization in nonphagocytic cells and a type III secretion system that allows antiphagocytic activity (Cornelis 2002; Isberg et al. 2000). These studies show that the border between intracellular and extracellular pathogens is somewhat artificial and even though some bacteria are predominantly located extracellularly, intracellular steps are key in bacterial survival and pathogenesis. Even if multiplication is extracellular, an intense dialogue takes place between these bacteria and host cells and adhering bacteria modulate various cellular responses controlled by small GTPases of the Rho family.

References

Allen LA, Schlesinger LS, Kang B (2000) Virulent strains of *Helicobacter pylori* demonstrate delayed phagocytosis and stimulate homotypic phagosome fusion in macrophages. J Exp Med 191:115–28

Barbieri AM, Sha Q, Bette-Bobillo P, Stahl PD, Vidal M (2001) ADP-ribosylation of Rab5 by ExoS of *Pseudomonas aeruginosa* affects endocytosis. Infect Immun 69:5329–34

Ben-Ami G, Ozeri V, Hanski E, Hofmann F, Aktories K, Hahn KM, Bokoch GM, Rosenshine I (1998) Agents that inhibit Rho, Rac, and Cdc42 do not block formation of actin pedestals in HeLa cells infected with enteropathogenic *Escherichia coli*. Infect Immun 66:1755–8

Bodger K, Crabtree JE (1998) *Helicobacter pylori* and gastric inflammation. Br Med Bull 54:139–50

Campellone KG, Leong JM (2003) Tails of two Tirs: actin pedestal formation by enteropathogenic *E. coli* and enterohemorrhagic *E. coli* O157:H7. Curr Opin Microbiol 6:82–90

Caron E, Hall A (1998) Identification of two distinct mechanisms of phagocytosis controlled by different Rho GTPases. Science 282:1717–21

Chimini G, Chavrier P (2000) Function of Rho family proteins in actin dynamics during phagocytosis and engulfment. Nat Cell Biol 2: E191–6

Churin Y, Al-Ghoul L, Kepp O, Meyer TF, Birchmeier W, Naumann M (2003) *Helicobacter pylori* CagA protein targets the c-Met receptor and enhances the motogenic response. J Cell Biol 161:249–55

Churin Y, Kardalinou E, Meyer TF, Naumann M (2001) Pathogenicity island-dependent activation of Rho GTPases Rac1 and Cdc42 in *Helicobacter pylori* infection. Mol Microbiol 40:815–23

Cornelis GR (2002) The *Yersinia* Ysc-Yop 'type III' weaponry. Nat Rev Mol Cell Biol 3:742–52

Crawford HC, Scoggins CR, Washington MK, Matrisian LM, Leach SD (2002) Matrix metalloproteinase-7 is expressed by pancreatic cancer precursors and regulates acinar-to-ductal metaplasia in exocrine pancreas. J Clin Invest 109:1437–44

Donnenberg MS, Donohue-Rolfe A, Keusch GT (1989) Epithelial cell invasion: an overlooked property of enteropathogenic *Escherichia coli* (EPEC) associated with the EPEC adherence factor. J Infect Dis 160:452–9

Donnenberg MS, Kaper JB (1992) Enteropathogenic *Escherichia coli*. Infect Immun 60:3953–61

Ernst JD (2000) Bacterial inhibition of phagocytosis. Cell Microbiol 2:379–86

Etienne-Manneville S, Hall A (2002) Rho GTPases in cell biology. Nature 420:629–35

Eugene E, Hoffmann I, Pujol C, Couraud PO, Bourdoulous S, Nassif X (2002) Microvilli-like structures are associated with the internalization of virulent capsulated *Neisseria meningitidis* into vascular endothelial cells. J Cell Sci 115:1231–41

Fleiszig SM, Evans DJ, Do N, Vallas V, Shin S, Mostov KE (1997) Epithelial cell polarity affects susceptibility to *Pseudomonas aeruginosa* invasion and cytotoxicity. Infect Immun 65:2861–7

Ganesan AK, Frank DW, Misra RP, Schmidt G, Barbieri JT (1998) *Pseudomonas aeruginosa* exoenzyme S ADP-ribosylates Ras at multiple sites. J Biol Chem 273:7332–7

Geiser TK, Kazmierczak BI, Garrity-Ryan LK, Matthay MA, Engel JN (2001) *Pseudomonas aeruginosa* ExoT inhibits in vitro lung epithelial wound repair. Cell Microbiol 3:223–36

Glocker E, Lange C, Covacci A, Bereswill S, Kist M, Pahl HL (1998) Proteins encoded by the cag pathogenicity island of *Pseudomonas aeruginosa* are required for NF-κB activation. Infect Immun 66: 2346–8

Guillemin K, Salama NR, Tompkins LS, Falkow S (2002) Cag pathogenicity island-specific responses of gastric epithelial cells to *Helicobacter pylori* infection. Proc Natl Acad Sci USA 99:15136–41

Hoffmann I, Eugene E, Nassif X, Couraud PO, Bourdoulous S (2001) Activation of ErbB2 receptor tyrosine kinase supports invasion of endothelial cells by *Neisseria meningitidis*. J Cell Biol 155:133–43

Isberg RR, Hamburger Z, Dersch P (2000) Signaling and invasin-promoted uptake via integrin receptors. Microbes Infect 2:793–801

Jepson MA, Pellegrin S, Peto L, Banbury DN, Leard AD, Mellor H, Kenny B (2003) Synergistic roles for the Map and Tir effector molecules in mediating uptake of enteropathogenic *Escherichia coli* (EPEC) into non-phagocytic cells. Cell Microbiol 5:773–83

Kallstrom H, Hansson-Palo P, Jonsson AB (2000) Cholera toxin and extracellular Ca^{2+} induce adherence of non-piliated *Neisseria*: evidence for an important role of G-proteins and Rho in the bacteria-cell interaction. Cell Microbiol 2:341–51

Kallstrom H, Liszewski MK, Atkinson JP, Jonsson AB (1997) Membrane cofactor protein (MCP or CD46) is a cellular pilus receptor for pathogenic *Neisseria*. Mol Microbiol 25:639–47

Kazi JI, Sinniah R, Jaffrey NA, Alam SM, Zaman V, Zuberi SJ, Kazi AM (1989) Cellular and humoral immune responses in *Campylobacter pylori*-associated chronic gastritis. J Pathol 159:231–7

Kazmierczak BI, Engel JN (2002) *Pseudomonas aeruginosa* ExoT acts in vivo as a GTPase-activating protein for RhoA, Rac1, and Cdc42. Infect Immun 70:2198–205

Kazmierczak BI, Mostov K, Engel JN (2004) Epithelial cell polarity alters Rho-GTPase responses to *Pseudomonas aeruginosa*. Mol Biol Cell 15:411–9

Kenny B, DeVinney R, Stein M, Reinscheid DJ, Frey EA, Finlay BB (1997) Enteropathogenic *E. coli* (EPEC) transfers its receptor for intimate adherence into mammalian cells. Cell 91:511–20

Kenny B, Ellis S, Leard AD, Warawa J, Mellor H, Jepson MA (2002) Co-ordinate regulation of distinct host cell signalling pathways by multifunctional enteropathogenic *Escherichia coli* effector molecules. Mol Microbiol 44:1095–1107

Kenny B, Jepson M (2000) Targeting of an enteropathogenic *Escherichia coli* (EPEC) effector protein to host mitochondria. Cell Microbiol 2:579–90

Krall R, Schmidt G, Aktories K, Barbieri JT (2000) *Pseudomonas aeruginosa* ExoT is a Rho GTPase-activating protein. Infect Immun 68:6066–8

Lommel S, Benesch S, Rottner K, Franz T, Wehland J, Kuhn R (2001) Actin pedestal formation by enteropathogenic *Escherichia coli* and intracellular motility of *Shigella flexneri* are abolished in N-WASP-defective cells. EMBO Rep 2:850–7

Martinez JJ, Hultgren SJ (2002) Requirement of Rho-family GTPases in the invasion of Type 1-piliated uropathogenic *Escherichia coli*. Cell Microbiol 4:19–28

Martinez JJ, Mulvey MA, Schilling JD, Pinkner JS, Hultgren SJ (2000) Type 1 pilus-mediated bacterial invasion of bladder epithelial cells. EMBO J 19:2803–12

Matsuda M, Tanaka S, Nagata S, Kojima A, Kurata T, Shibuya M (1992) Two species of human CRK cDNA encode proteins with distinct biological activities. Mol Cell Biol 12:3482–9

McDonnell S, Navre M, Coffey RJ, Jr., Matrisian LM (1991) Expression and localization of the matrix metalloproteinase pump-1 (MMP-7) in human gastric and colon carcinomas. Mol Carcinog 4:527–33

Merz AJ, Enns CA, So M (1999) Type IV pili of pathogenic *Neisseriae* elicit cortical plaque formation in epithelial cells. Mol Microbiol 32:1316–32

Mulvey MA, Lopez-Boado YS, Wilson CL, Roth R, Parks WC, Heuser J, Hultgren SJ (1998) Induction and evasion of host defenses by type 1-piliated uropathogenic *Escherichia coli*. Science 282:1494–7

Mulvey MA, Schilling JD, Hultgren SJ (2001) Establishment of a persistent *Escherichia coli* reservoir during the acute phase of a bladder infection. Infect Immun 69:4572–9

Nassif X, Bourdoulous S, Eugene E, Couraud PO (2002) How do extracellular pathogens cross the blood-brain barrier? Trends Microbiol 10:227–32

Naumann M, Wessler S, Bartsch C, Wieland B, Covacci A, Haas R, Meyer TF (1999) Activation of activator protein 1 and stress response kinases in epithelial cells colonized by *Helicobacter pylori* encoding the cag pathogenicity island. J Biol Chem 274:31655–62

Phillips AD, Giron J, Hicks S, Dougan G, Frankel G (2000) Intimin from enteropathogenic *Escherichia coli* mediates remodelling of the eukaryotic cell surface. Microbiology 146 (Pt 6): 1333–44

Raftopoulou M, Hall A (2004) Cell migration: Rho GTPases lead the way. Dev Biol 265:23–32

Ramarao N, Gray-Owen SD, Backert S, Meyer TF (2000) *Helicobacter pylori* inhibits phagocytosis by professional phagocytes involving type IV secretion components. Mol Microbiol 37:1389–404

Rayner CF, Dewar A, Moxon ER, Virji M, Wilson R (1995) The effect of variations in the expression of pili on the interaction of *Neisseria meningitidis* with human nasopharyngeal epithelium. J Infect Dis 171:113–21

Ren XD, Kiosses WB, Schwartz MA (1999) Regulation of the small GTP-binding protein Rho by cell adhesion and the cytoskeleton. EMBO J 18:578–85

Rosenshine I, Ruschkowski S, Stein M, Reinscheid DJ, Mills SD, Finlay BB (1996) A pathogenic bacterium triggers epithelial signals to form a functional bacterial receptor that mediates actin pseudopod formation. EMBO J 15:2613–24

Sauer FG, Mulvey MA, Schilling JD, Martinez JJ, Hultgren SJ (2000) Bacterial pili: molecular mechanisms of pathogenesis. Curr Opin Microbiol 3:65–72

Segal ED, Cha J, Lo J, Falkow S, Tompkins LS (1999) Altered states: involvement of phosphorylated CagA in the induction of host cellular growth changes by *Helicobacter pylori*. Proc Natl Acad Sci USA 96:14559–64

Selbach M, Moese S, Hurwitz R, Hauck CR, Meyer TF, Backert S (2003) The *Helicobacter pylori* CagA protein induces cortactin dephosphorylation and actin rearrangement by c-Src inactivation. EMBO J 22:515–28

Sim RJ, Harrison MM, Moxon ER, Tang CM (2000) Underestimation of meningococci in tonsillar tissue by nasopharyngeal swabbing. Lancet 356:1653–4

Stephens DS, Hoffman LH, McGee ZA (1983) Interaction of *Neisseria meningitidis* with human nasopharyngeal mucosa: attachment and entry into columnar epithelial cells. J Infect Dis 148:369–76

Sun J, Barbieri JT (2003) *Pseudomonas aeruginosa* ExoT ADP-ribosylates CT10 regulator of kinase (Crk) proteins. J Biol Chem 278:32794–800

Symons M, Rusk N (2003) Control of vesicular trafficking by Rho GTPases. Curr Biol 13: R409–18

Thomas WE, Trintchina E, Forero M, Vogel V, Sokurenko EV (2002) Bacterial adhesion to target cells enhanced by shear force. Cell 109:913–23

Tsutsumi R, Higashi H, Higuchi M, Okada M, Hatakeyama M (2003) Attenuation of *Helicobacter pylori* CagA x SHP-2 signaling by interaction between CagA and C-terminal Src kinase. J Biol Chem 278:3664–70

Vallance BA, Finlay BB (2000) Exploitation of host cells by enteropathogenic *Escherichia coli*. Proc Natl Acad Sci USA 97:8799–806

Van Aelst L, Symons M (2002) Role of Rho family GTPases in epithelial morphogenesis. Genes Dev 16:1032–54

Wroblewski LE, Noble PJ, Pagliocca A, Pritchard DM, Hart CA, Campbell F, Dodson AR, Dockray GJ, Varro A (2003) Stimulation of MMP-7 (matrilysin) by *Helicobacter pylori* in human gastric epithelial cells: role in epithelial cell migration. J Cell Sci 116:3017–26

Yahr TL, Barbieri JT, Frank DW (1996) Genetic relationship between the 53- and 49-kilodalton forms of exoenzyme S from *Pseudomonas aeruginosa*. J Bacteriol 178:1412–9

CTMI (2005) 291:29–42

Triggered Phagocytosis by *Salmonella*: Bacterial Molecular Mimicry of RhoGTPase Activation/Deactivation

M. C. Schlumberger · W.-D. Hardt (✉)

Institute of Microbiology, ETH Zürich, Wolfgang-Pauli-Strasse 10, 8093 Zürich, Switzerland
hardt@micro.biol.ethz.ch

Abstract *Salmonella* Typhimurium uses the type III secretion system encoded in the *Salmonella* pathogenicity island I (SPI-1 TTSS) to inject toxins (effector proteins) into host cells. Here, we focus on the functional mechanism of three of these toxins: SopE, SopE2, and SptP. All three effector proteins change the GTP/GDP loading state of RhoGTPases by transient interactions. SopE and SopE2 mimic eukaryotic G-nucleotide exchange factors and thereby activate RhoGTPase signaling pathways in infected host cells. In contrast, a domain of SptP inactivates RhoGTPases by mimicking the activity of eukaryotic GTPase-activating proteins. The *Salmonella*-host cell interaction provides an excellent example for the use of molecular mimicry by bacterial pathogens.

1
Host Cell Invasion via the Trigger Mechanism: The *Salmonella* Paradigm

Salmonella spp. are gram-negative rods that cause foodborne infections, worldwide. After ingestion of contaminated food, the bacteria reach the gut, where they trigger diarrhea and invade the intestinal mucosa. The latter process can be simulated in tissue culture: Five to fifteen minutes after addition of *Salmonella enterica* subspecies 1 serovar Typhimurium (*S.* Typhimurium) to fibroblasts (and many other nonphagocytic cell lines) the bacteria induce profound cytoskeletal rearrangements in the infected mammalian cell. These actin-rich membrane ruffles engulf the bacteria and lead to *Salmonella* invasion into the host cell. Soon after the discovery of this phenomenon it was shown that *S.* Typhimurium uses a specific microinjection organelle called the SPI-1 type III secretion system (SPI-1 TTSS) to induce these responses (Fig. 1) [14]. This has set the stage for the molecular analysis of *Salmonella* host cell invasion, which has led to a series of breakthrough discoveries in the past decade.

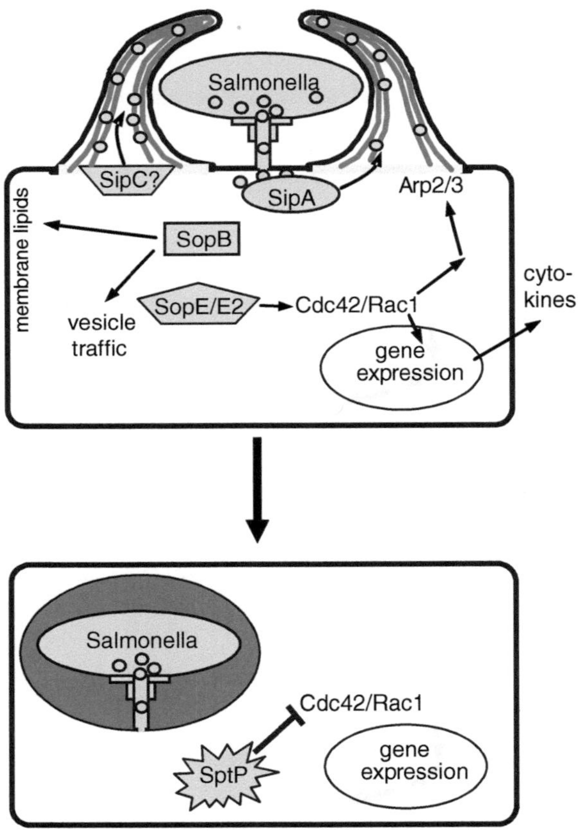

Fig. 1 Triggering of host cell invasion by the *S.* Typhimurium SPI-1 TTSS

Two strategies have been of key importance:

a. The analysis of the signaling pathways that are involved in *Salmonella* host cell invasion. In a breakthrough study it was found that the RhoGTPases Cdc42 and Rac1 are key players in this process [6].

b. Analysis of the toxins (termed "effector proteins") that are transported via the SPI-1 TTSS. In the past 7 years, 12 effector proteins have been identified that travel via the SPI-1 TTSS [13]. Construction of *Salmonella* mutants that lack one or more components of the effector protein cocktail has identified 4 SPI-1 effectors as key players in the triggering of host cell invasion: SipA, SopB, SopE
indexEffector protein!SopE, SopE2. A *sipAsopBEE2 S.* Typhimurium mutant is practically noninvasive [9, 27, 42]. A fifth effector protein, SptP, has been found to reverse actin cytoskeletal rearrangements once the bacteria have arrived inside the host cell [11, 12]. Other effector proteins might also be involved in manipulation of the host cell actin cytoskeleton (i.e., SipC [19]). However, their exact role is still a matter of dispute.

In recent years biochemical studies, structural analyses, transfection, and microinjection assays have provided rich insights into the molecular basis of SipA, SopB, SopE2, and SopE function.

SipA regulates actin cytoskeletal dynamics at the level of the actin filament by two types of mechanism: It binds to and stabilizes actin filaments in a nebulin-like fashion [15, 20, 21, 24, 26, 28, 43, 44]. Furthermore, *Salmonella* spp. interfere with ADF/cofilin and gelsolin function, two key F-actin destabilizing proteins, either directly via SipA or indirectly by triggering Cdc42-dependent signaling cascades [8, 25].

SopB is a phosphatidylinositol phosphatase [32]. Currently, two different mechanisms are discussed to explain how SopB triggers host cell invasion: SopB might reduce PtdIns(4,5)P_2 levels in the ruffling membrane and thereby promote the rapid formation of *Salmonella*-containing phagosomes [40]. Other work indicates that SopB can also enhance actin cytoskeleton rearrangements and bacterial entry by triggering a signaling pathway involving Cdc42 [42]. Further work is required to determine exactly how manipulation of phospholipid levels activates RhoGTPases during *Salmonella* host cell invasion. For more detailed information, the reader is referred to two excellent recent reviews [2, 3].

SopE and SopE2 are G-nucleotide exchange factors (GEF) for host cell RhoGTPases and trigger host cell invasion by direct activation of Cdc42 and Rac1 [17, 37]. SptP, on the other hand, inactivates RhoGTPase signaling. It is a bifunctional enzyme with a tyrosine phosphatase- and a GTPase activation

(GAP) domain [11, 12, 22]. The GEF and GAP domains of SopE, SopE2, and SptP are excellent examples for "molecular mimicry"—bacterial virulence factors imitating enzymes of the host cell. They will be discussed in detail in this review.

2
The SopE Protein Family

SopE was discovered as a protein secreted by *S.* Dublin and *S.* Typhimurium via the SPI-1 TTSS [18, 41]. Later, it was found that *Salmonella* spp. often encode two SopE-like proteins: SopE and SopE2 [1, 37]. Interestingly, *sopE2* genes are present in all and *sopE* only in some *Salmonella* spp. [27, 37]. In the *sopE*$^+$ strains, SopE is encoded in prophages or horizontally acquired "pathogenicity islets." Both SopE and SopE2 are 240 aa in size, are about 70% identical, and are transported via the SPI-1 TTSS. Because SopE has been studied in much more detail than SopE2, this review will focus mostly on SopE. SopE is present in the *S.* Typhimurium strain SL1344 [39], which is widely used to study *Salmonella* pathogenesis.

The protein sequence did not yield much information about the possible function of SopE proteins: It did not show any sequence similarity to any other known eukaryotic or prokaryotic proteins. Recently, however, a third distantly related member (BopE) of the SopE protein family has been discovered in *Burkholderia* spp. [38]. First hints about the involvement of SopE proteins in host cell manipulation originated from mutation analyses. SopE mutants were slightly but significantly less invasive for tissue culture cells than the wild-type strains [18, 41]. Later it became clear that SopE, SopE2, SopB, and SipA all participate in the triggering of host cell invasion. In the background of *sopE2sopBsipA* strains, disruption of *sopE* has a very pronounced invasion phenotype [9, 16, 27, 42].

3
Molecular Function of SopE

First insights into the molecular function of SopE were obtained in a screen for SopE interacting proteins in a HeLa cDNA library [17]. This screen identified Rac1 and Cdc42 as SopE interaction partners. In vitro assays demonstrated that SopE mimicks the activity of G-nucleotide exchange factors of eukaryotic cells: It efficiently binds to "inactive" GDP*RhoGTPase complexes, thereby displacing the bound G-nucleotide. This step of the catalytic cycle has been

Table 1 Kinetic parameters of SopE-mediated G-nucleotide exchange on recombinant, non-farnesylated Cdc42

Michaelis–Menten kinetics*	
k_{cat}	1 s^{-1}
K_m	2–4 μM
k_{cat}/K_m	2–6 × 10^5 $M^{-1}s^{-1}$
Single turnover kinetics (surplus enzyme)	
k_{react}	>13 s^{-1}
K_m pre-steady state	>10 μM
k_{react}/K_m pre-steady state	7 × 10^5 M^{-1} s^{-1}
Surface plasmon resonance	
k_{ass} (no GDP)	6 × 10^5 M^{-1} s^{-1}
k_{off} (no GDP)	1.4 × 10^{-4} s^{-1}
$K_D=k_{off}/k_{ass}$ (no GDP)	~ 1–10 nM
k_{off} (1 mM GDP)	0.6 s^{-1}

Determined at 25 °C [33, 34].
*uncatalyzed GDP-release reaction: $k_{cat} = 5 \times 10^5$ s^{-1} [34]

studied in detail by surface plasmon resonance analyses (Table 1) [10, 33]). In the presence of free GDP or GTP, the SopE*RhoGTPase complex is unstable and the SopE is quickly displaced by free GDP or GTP (Table 1). Thus the overall reaction cycle results in a drastically increased G-nucleotide exchange rate on the RhoGTPase. In a mammalian cell, the GDP-to-GTP ratio is generally in the range of 1:10 Kinetic analyses revealed that SopE accelerates the net conversion of "inactive" GDP*RhoGTPase complexes into the "active" GTP*RhoGTPase conformation by about 10^5-fold (Table 1). Interestingly, the catalytic rate (k_{cat}) of SopE is about 10- to 100-fold higher than that of the catalytic domains of eukaryotic GEFs. One might speculate that this is of importance in the early phase of the bacteria-host cell interaction when only little effector protein has been transported into the host cell. A high catalytic rate (more specifically a high catalytic efficiency (k_{cat}/K_m) would ensure activation of a significant fraction of the host cellular RhoGTPase pool shortly after host cell contact.

4 Structure of the SopE*Cdc42 Complex: Molecular Mimicry

The nucleotide-free complex between the catalytic domain of SopE (aa 78–240) and a truncated version of Cdc42 (aa 1–178) lacking the C-terminal prenylation site has been crystallized, and the structure of this complex was determined at 2.3-Å resolution [4].

Table 2 Characteristics of site-specific SopE mutants (reprinted with permission from [34])

SopE variant	Ruffling* ++ (%)	Ruffling + (%)	Ruffling – (%)	Number of cells analyzed	Catalytic performance (% of wt)[a]
No SopE	0	3	**97**	471	
wt	**81**	16	2	443	100
Q109N	14	**75**	11	352	0.6
Q109A	1	39	**60**	296	0.1
D124E	10	**87**	3	270	0.2
D124A	2	**82**	15	241	0.2
K198A	**91**	9	0	395	60
K198R	**92**	7	1	318	89
K198E	**75**	25	0	226	2.8
D103E	**92**	7	1	291	42
D103A	**86**	13	1	280	21
delG168	0	24	**76**	153	0.1
G168A	0	16	**84**	215	0.02
G168V	0	12	**88**	237	0.004

[a]Single-turnover catalytic efficiencies expressed in % of the efficiencies for SopE′ wt or for GST-SopE′ wt

*Ability of SopE variants to indice cytoskeletal rearrangements in cultured cells

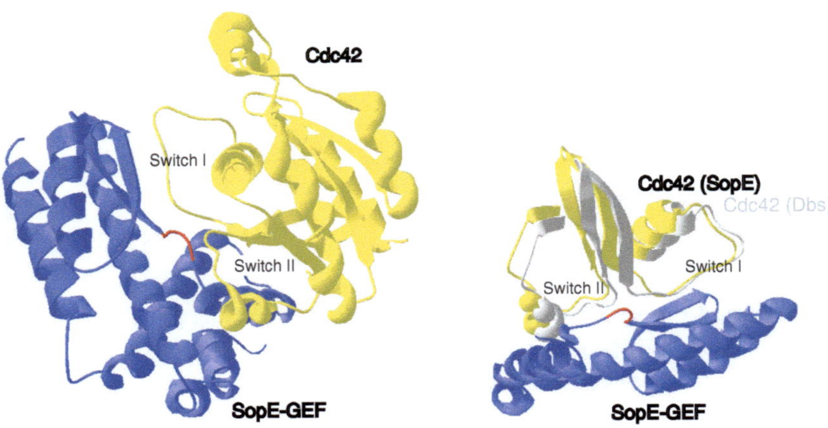

Fig. 2 Structure of the SopE Cdc42 complex: molecular mimicry of eukaryotic GEFs. The catalytic GAGA-loop of SopE is shown in *red*

The catalytic domain of SopE is almost entirely α-helical. It forms two three-helix bundles that are arranged in a V-like fashion. A small loop (^{166}GAGA^{169}) is located at the junction between these two domains. This ^{166}GAGA^{169} loop represents a large part of the interaction surface with Cdc42, and mutagenesis studies demonstrated that the ^{166}GAGA^{169} loop is of great importance for SopE function (Fig. 2; Table 2). The ^{166}GAGA^{169} loop inserts between and displaces the switch 1 and switch 2 regions of Cdc42. Switch 1 and switch 2 are essential components of the G-nucleotide binding pocket of Cdc42. Thus SopE enhances G-nucleotide release by distorting the switch 1 and switch 2 regions of Cdc42.

The structures of several complexes between RhoGTPases and eukaryotic GEFs (Tiam1, Dbs, intersectin) are also available. These eukaryotic GEFs have structures entirely different from SopE. However, they bind to the same site of their cognate RhoGTPase (switches 1 and 2) and, most importantly, they distort switch 1 and switch 2 into virtually the same conformation as observed in the SopE*Cdc42 complex (Fig. 2). Thus SopE accurately mimics the catalytic strategy of eukaryotic G-nucleotide exchange factors to activate RhoGTPase signaling.

5 G-Nucleotide Exchange Catalyzed by SopE Versus EDTA

It is quite interesting to observe that SopE$_{78-240}$-catalyzed G-nucleotide release can proceed at higher rates (k_{cat}) than G-nucleotide release in the presence of EDTA [4, 33, 37]. The latter reaction is commonly used for loading G-binding proteins with new nucleotides or as a control for G-nucleotide release experiments.

The different G-release rates obtained with a Mg^{2+} chelator (EDTA) and SopE can be rationalized if one considers the underlying catalytic mechanisms. These are quite different, even though the net result (dissociation of a G-nucleotide that was formerly bound to the G protein) is the same.

In general, RhoGTPases bind the G-nucleotide with three regions of the protein (switch I, switch II, P-loop). In addition, the RhoGTPases bind a magnesium ion, which stabilizes the phosphate of the G-nucleotide. This Mg^{2+} interaction is required for high-affinity G-nucleotide binding. EDTA disrupts this stabilizing interaction by removing the magnesium ion. SopE uses an entirely different mechanism: SopE (very much like eukaryotic GEFs for RhoGTPases) does not bind directly to either the G-nucleotide or the magnesium ion. Instead, it binds to and distorts the switch I and II regions of the RhoGTPase into a conformation that disrupts the binding of switch I and II to the G-nucleotide and also the Mg^{2+} binding site of the RhoGTPase. SopE binds very fast and very efficiently to the RhoGTPase, and this explains why G-nucleotide exchange at high SopE concentrations (k_{cat}) is much faster than the rates observed in the presence of EDTA.

6 Signaling of SopE Proteins Inside the Host Cell

Transfection of SopE expression vectors or microinjection of SopE protein into fibroblasts leads to profound cytoskeletal rearrangements and cellular uptake of noninvasive bacteria [17]. Furthermore, cotransfection experiments showed that SopE expression leads to activation of the Jun-kinase pathway [5, 17]. Both responses are suppressed by dominant negative versions of Cdc42 and Rac1. This demonstrated that SopE is sufficient to activate Cdc42 and Rac1 inside eukaryotic cells. Later, pull-down assays with CRIB-domain constructs have verified that Cdc42 and Rac1 are also activated by SopE during the course of a bacterial infection [10].

Many *Salmonella* spp. including the strain S. Typhimurium SL1344 harbor *sopE* and *sopE2*. It was found that SopE and SopE2 can mediate efficient host cell invasion and that both are efficient GEFs for Cdc42 [37]. SopE is also an efficient GEF for Rac1, whereas SopE2 is not [10]. This was also confirmed in COS-7 tissue culture infection experiments. These observations suggest that subtle differences in substrate specificity might allow certain *Salmonella* spp. expressing both SopE and SopE2 in parallel to optimize their interaction with host cells.

At present it is unclear which other host cell GTPases (besides Rac1 and Cdc42) might also be manipulated by SopE or SopE2 in vivo. Only few RhoGTPases have been tested as SopE substrates in vitro [17]. One group reported that SopE proteins might also activate RabGTPases involved in vesicular trafficking [29]. Furthermore, the requirement of Cdc42 and/or Rac1 activation by S. Typhimurium might vary significantly between different cell types and even between the two surfaces of the same polarized cell line [7, 8].

7 Identification and Domain Structure of SptP

SptP was discovered during sequence analysis of the S. Typhimurium locus encoding the SPI-1 TTSS [22]. The 60-kDa SptP protein has three domains: an N-terminal secretion domain that is required for binding to a specific transport chaperone (SicP [35]) and transport via the SPI-1 TTSS, a central domain that is a GAP for host cell RhoGTPases (aa 174–290; discussed below), and a C-terminal tyrosine phosphatase domain (aa 340–543). The tyrosine phosphatase domain of SptP is similar to tyrosine phosphatase domains from other bacterial type III effector proteins (i.e., YopH from *Yersinia* spp.) and includes a conserved PTPase signature sequence (LIVMF)-H-C-X(2)-G-X(3)-(STC)-(STAG)-X-(LIVMFY) Disruption of the conserved Cys479 abolishes the tyrosine phosphatase activity of SptP [22]. The tyrosine phosphatase domain of SptP has a canonical phosphatase fold [35]. Interestingly, tyrosine phosphatases are absent from most bacteria. Therefore, the few bacterial tyrosine phosphatases that exist are thought to originate from some event of horizontal gene transfer from a eukaryote. In the host cell, SptP can modulate MAP kinase signaling and modulate the structure of the actin cytoskeleton [12, 30]. So far, no direct effect of the tyrosine phosphatase domain of SptP on RhoGTPases has been observed. Therefore, we will focus our discussion of SptP on the GAP domain.

8 The GAP Domain of SptP

The GAP Domain of SptP is similar to the type III effector proteins ExoS from *Pseudomonas* spp. and YopE from *Yersinia* spp. [22]. Most importantly, these proteins share a conserved arginine residue. An arginine residue has been known to act as the catalytic center of eukaryotic GAP proteins (see below).

In vitro the GAP domain of SptP is a potent activator of the GTPase activity of RhoGTPases, namely Rac1 and Cdc42, and preferentially binds to their GTP-bound forms [11]. Exchange of the conserved arginine residue of SptP (R209A) resulted in loss of GAP activity [11]. Several lines of evidence confirmed that the GAP domain of SptP is a key modulator of host cell responses: Microinjection of SptP resulted in disruption of cytoskeletal structures and reversal of F-actin rearrangements induced by other *Salmonella* effector proteins (namely SopE and SopE2); It could reverse the activation of JNK signaling that is induced during the *Salmonella* host cell interaction [11].

9 Structure of the SptP-GAP*Rac1 Complex

The structure of a fragment of SptP containing the GAP and tyrosine phosphatase domains (aa 161-543) has been determined (Fig. 3) [36]. The GAP-domain forms a four-helix bundle with supercoiled helices, and its fold is quite

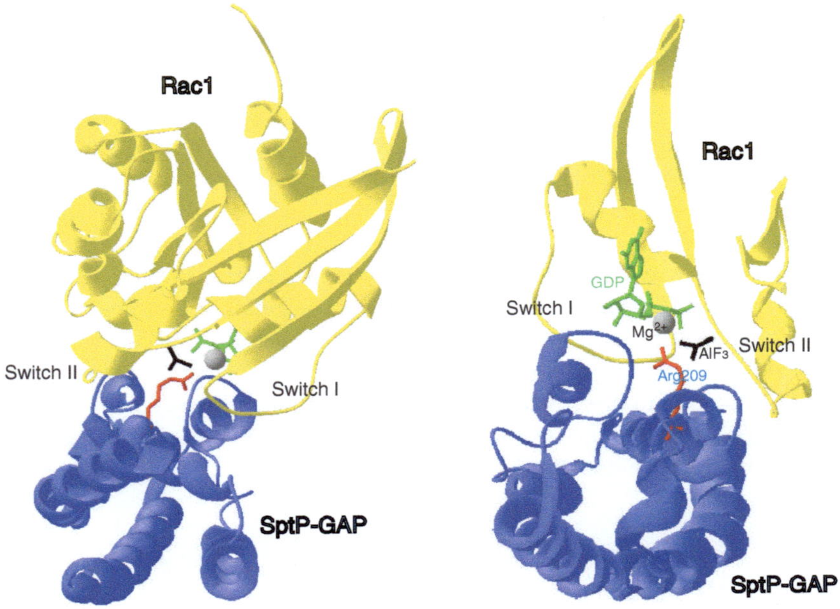

Fig. 3 Structure of the SptP(GAP)-Rac1 complex: molecular mimicry of eukaryotic GAPs. The arginine finger of SptP(GAP) is shown in *red*

different from that of eukaryotic GAPs. The structure of the SptP-GAP*Rac1 transition-state complex with GDP and aluminum fluoride has also been solved [36]. In this complex, the AlF_3 occupies the site of the γ-phosphate of GTP. The SptP-GAP helices H1 and H4 as well as one bulged-out portion of the GAP domain make the contacts with the RhoGTPase. They contact key elements in the Rac1 nucleotide binding pocket (Gln61 of switch II, aluminum fluoride in the nucleotide binding pocket, GDP). The switch I region of Rac1, which is in intimate contact with SptP-GAP, is forced into a highly ordered conformation similar to that found in complexes between eukaryotic GAPs and their cognate RhoGTPases.

The catalytically important arginine residue R209 at the end of helix H4 extends deeply into the "active site cavity" of Rac1. The structure implies that its guanidinium group forms strong hydrogen bonds with the terminal GTP phosphates, "complements" the otherwise suboptimal GTPase center of RhoGTPases, and thereby drastically accelerates hydrolysis of the γ-phosphate of GTP. The same mechanism has been observed in eukaryotic GAPs, i.e., the complex between Cdc42 and Cdc42GAP [31]. The term "arginine finger" has been coined for the common arginine residue at the catalytic center of GAPs.

In conclusion, the SptP-GAP has a distinct three-dimensional fold but a basic catalytic mechanism very similar to eukaryotic GAPs. In other words, the SptP-GAP domain has emerged by convergent evolution to mimic eukaryotic GAPs.

10
SopE Yin-SptP Yang

It has been quite puzzling to find two groups of effector proteins with seemingly opposite functions that are both transported in parallel via the same TTSS. Indeed, cotransfection of SopE and SptP expression vectors or microinjection of a mixture of both effector proteins into cultured cells alleviated the dramatic effects on the actin cytoskeleton and JNK signaling that are normally triggered by SopE [11]. Now it turns out that SopE and SptP might act in a sequential manner during *Salmonella*-host cell interaction. Both effector proteins are transported into the host cells early on. In the initial phase of the interaction, the activity of SopE seems to override the effect of SptP. The bacteria induce RhoGTPase activation, JNK activation, and dramatic cytoskeletal rearrangements that lead to entry of the bacteria into the host cells. Later, JNK activity comes back down and the cytoskeleton returns to normal. The latter phase is dominated by SptP. In a landmark study, it was demonstrated that this can be explained by distinct rates of degradation: SopE is degraded rapidly (within less than 1 h) in the host cell cytosol, whereas SptP is degraded slowly ($t_{1/2}$ >>1 h) [23]. Therefore, *S.* Typhimurium exploits a "clock mechanism" (i.e., ubiquitin-dependent degradation) of the host cell to coordinate the function of its effector proteins. The host proteasome determines the balance of the *Salmonella* yin and yang.

References

1. Bakshi, C. S., V. P. Singh, M. W. Wood, P. W. Jones, T. S. Wallis, and E. E. Galyov. 2000. Identification of SopE2, a *Salmonella* secreted protein which is highly homologous to SopE and involved in bacterial invasion of epithelial cells. J Bacteriol 182:2341–4.
2. Brumell, J. H., and S. Grinstein. 2003. Role of lipid-mediated signal transduction in bacterial internalization. Cell Microbiol 5:287–97.
3. Brumell, J. H., and S. Grinstein. 2004. *Salmonella* redirects phagosomal maturation. Curr Opin Microbiol 7:78–84.
4. Buchwald, G., A. Friebel, J. E. Galan, W. D. Hardt, A. Wittinghofer, and K. Scheffzek. 2002. Structural basis for the reversible activation of a Rho protein by the bacterial toxin SopE. EMBO J 21:3286–95.

5. Chen, L. M., S. Bagrodia, R. A. Cerione, and J. E. Galan. 1999. Requirement of p21-activated kinase (PAK) for *Salmonella* Typhimurium-induced nuclear responses. J Exp Med 189:1479–88.
6. Chen, L. M., S. Hobbie, and J. E. Galan. 1996. Requirement of CDC42 for *Salmonella*-induced cytoskeletal and nuclear responses. Science 274:2115–8.
7. Criss, A. K., D. M. Ahlgren, T. S. Jou, B. A. McCormick, and J. E. Casanova. 2001. The GTPase Rac1 selectively regulates *Salmonella* invasion at the apical plasma membrane of polarized epithelial cells. J Cell Sci 114:1331–41.
8. Criss, A. K., and J. E. Casanova. 2003. Coordinate regulation of *Salmonella enterica* serovar Typhimurium invasion of epithelial cells by the Arp2/3 complex and Rho GTPases. Infect Immun 71:2885–91.
8a. Dai, Sarmiere et al. 2004. Efficient Salmonella entry requires activity cycles of host ADF and cofilin. Cell Microbiol 6: 459–71.
9. Ehrbar, K., A. Friebel, S. I. Miller, and W. D. Hardt. 2003. Role of the *Salmonella* pathogenicity island 1 (SPI-1) protein InvB in type III secretion of SopE and SopE2, two *Salmonella* effector proteins encoded outside of SPI-1. J Bacteriol 185:6950–67.
10. Friebel, A., H. Ilchmann, M. Aelpfelbacher, K. Ehrbar, W. Machleidt, and W. D. Hardt. 2001. SopE and SopE2 from *Salmonella* Typhimurium activate different sets of RhoGTPases of the host cell. J Biol Chem 276:34035–34040.
11. Fu, Y., and J. E. Galan. 1999. A *Salmonella* protein antagonizes Rac-1 and Cdc42 to mediate host-cell recovery after bacterial invasion. Nature 401:293–7.
12. Fu, Y., and J. E. Galan. 1998. The *Salmonella typhimurium* tyrosine phosphatase SptP is translocated into host cells and disrupts the actin cytoskeleton. Mol Microbiol 27:359–68.
13. Galan, J. E. 2001. *Salmonella* interactions with host cells: type III secretion at work. Annu Rev Cell Dev Biol 17:53–86.
14. Galan, J. E., and R. Curtiss, 3rd. 1989. Cloning and molecular characterization of genes whose products allow *Salmonella typhimurium* to penetrate tissue culture cells. Proc Natl Acad Sci USA 86:6383–7.
15. Galkin, V. E., A. Orlova, M. S. VanLoock, D. Zhou, J. E. Galan, and E. H. Egelman. 2002. The bacterial protein SipA polymerizes G-actin and mimics muscle nebulin. Nat Struct Biol 9:518–21.
16. Hapfelmeier, S., K. Ehrbar, B. Stecher, M. Barthel, M. Kremer, and W. D. Hardt. 2004. Role of the salmonella pathogenicity island 1 effector proteins SipA, SopB, SopE, and SopE2 in *Salmonella enterica* subspecies 1 serovar Typhimurium colitis in streptomycin-pretreated mice. Infect Immun 72:795–809.
17. Hardt, W. D., L. M. Chen, K. E. Schuebel, X. R. Bustelo, and J. E. Galan. 1998. *S. typhimurium* encodes an activator of Rho GTPases that induces membrane ruffling and nuclear responses in host cells. Cell 93:815–26.
18. Hardt, W. D., H. Urlaub, and J. E. Galan. 1998. A substrate of the centisome 63 type III protein secretion system of *Salmonella typhimurium* is encoded by a cryptic bacteriophage. Proc Natl Acad Sci USA 95:2574–9.
19. Hayward, R. D., and V. Koronakis. 1999. Direct nucleation and bundling of actin by the SipC protein of invasive *Salmonella*. EMBO J 18:4926–34.
20. Higashide, W., S. Dai, V. P. Hombs, and D. Zhou. 2002. Involvement of SipA in modulating actin dynamics during *Salmonella* invasion into cultured epithelial cells. Cell Microbiol 4:357–65.
21. Jepson, M. A., B. Kenny, and A. D. Leard. 2001. Role of sipA in the early stages of *Salmonella* Typhimurium entry into epithelial cells. Cell Microbiol 3:417–26.

22. Kaniga, K., J. Uralil, J. B. Bliska, and J. E. Galan. 1996. A secreted protein tyrosine phosphatase with modular effector domains in the bacterial pathogen *Salmonella typhimurium*. Mol Microbiol 21:633–41.
23. Kubori, T., and J. E. Galan. 2003. Temporal regulation of salmonella virulence effector function by proteasome-dependent protein degradation. Cell 115:333–42.
24. Lilic, M., V. E. Galkin, A. Orlova, M. S. VanLoock, E. H. Egelman, and C. E. Stebbins. 2003. *Salmonella* SipA polymerizes actin by stapling filaments with nonglobular protein arms. Science 301:1918–21.
25. McGhie, E. J., R. D. Hayward, and V. Koronakis. 2004. Control of actin turnover by a salmonella invasion protein. Mol Cell 13:497–510.
26. McGhie, E. J., R. D. Hayward, and V. Koronakis. 2001. Cooperation between actin-binding proteins of invasive *Salmonella*: SipA potentiates SipC nucleation and bundling of actin. EMBO J 20:2131–9.
27. Mirold, S., K. Ehrbar, A. Weissmüller, R. Prager, H. Tschäpe, H. Rüssmann, and W. D. Hardt. 2001. *Salmonella* host cell invasion emerged by acquisition of a mosaic of separate genetic elements, including *Salmonella* pathogenicity island 1 (SPI1), SPI5, and *sopE2*. J Bacteriol 183:2348–2358.
28. Mitra, K., D. Zhou, and J. E Galan. 2000. Biophysical characterization of SipA, an actin-binding protein from *Salmonella enterica*. FEBS Lett 482:81–4.
29. Mukherjee, K., S. Parashuraman, M. Raje, and A. Mukhopadhyay. 2001. SopE acts as an Rab5-specific nucleotide exchange factor and recruits non-prenylated Rab5 on *Salmonella*-containing phagosomes to promote fusion with early endosomes. J Biol Chem 276:23607–15.
30. Murli, S., R. O. Watson, and J. E. Galan. 2001. Role of tyrosine kinases and the tyrosine phosphatase SptP in the interaction of *Salmonella* with host cells. Cell Microbiol 3:795–810.
31. Nassar, N., G. R. Hoffman, D. Manor, J. C. Clardy, and R. A. Cerione. 1998. Structures of Cdc42 bound to the active and catalytically compromised forms of Cdc42GAP. Nat Struct Biol 5:1047–52.
32. Norris, F. A., M. P. Wilson, T. S. Wallis, E. E. Galyov, and P. W. Majerus. 1998. SopB, a protein required for virulence of *Salmonella dublin*, is an inositol phosphate phosphatase. Proc Natl Acad Sci USA 95:14057–9.
33. Rudolph, M. G., C. Weise, S. Mirold, B. Hillenbrand, B. Bader, A. Wittinghofer, and W. D. Hardt. 1999. Biochemical analysis of SopE from *Salmonella typhimurium*, a highly efficient guanosine nucleotide exchange factor for RhoGTPases. J Biol Chem 274:30501–9.
34. Schlumberger, M. C., A. Friebel, G. Buchwald, K. Scheffzek, A. Wittinghofer, and W. D. Hardt. 2003. Amino acids of the bacterial toxin SopE involved in G-nucleotide exchange on Cdc42. J Biol Chem.
35. Stebbins, C. E., and J. E. Galan. 2001. Maintenance of an unfolded polypeptide by a cognate chaperone in bacterial type III secretion. Nature 414:77–81.
36. Stebbins, C. E., and J. E. Galan. 2000. Modulation of host signaling by a bacterial mimic: structure of the *Salmonella* effector SptP bound to Rac1. Mol Cell 6:1449–60.
37. Stender, S., A. Friebel, S. Linder, M. Rohde, S. Mirold, and W. D. Hardt. 2000. Identification of SopE2 from *Salmonella typhimurium*, a conserved guanine nucleotide exchange factor for Cdc42 of the host cell. Mol Microbiol 36:1206–21.
38. Stevens, M. P., A. Friebel, L. A. Taylor, M. W. Wood, P. J. Brown, W. D. Hardt, and E. E. Galyov. 2003. A Burkholderia pseudomallei type III secreted protein, BopE, facilitates bacterial invasion of epithelial cells and exhibits guanine nucleotide exchange factor activity. J Bacteriol 185:4992–6.

39. Stocker, B. A., S. K. Hoiseth, and B. P. Smith. 1983. Aromatic-dependent *Salmonella* spp. as live vaccine in mice and calves. Dev Biol Stand 53:47–54.
40. Terebiznik, M. R., O. V. Vieira, S. L. Marcus, A. Slade, C. M. Yip, W. S. Trimble, T. Meyer, B. B. Finlay, and S. Grinstein. 2002. Elimination of host cell PtdIns(4,5)P(2) by bacterial SigD promotes membrane fission during invasion by *Salmonella*. Nat Cell Biol 4:766–73.
41. Wood, M. W., R. Rosqvist, P. B. Mullan, M. H. Edwards, and E. E. Galyov. 1996. SopE, a secreted protein of *Salmonella dublin*, is translocated into the target eukaryotic cell via a sip-dependent mechanism and promotes bacterial entry. Mol Microbiol 22:327–38.
42. Zhou, D., L. M. Chen, L. Hernandez, S. B. Shears, and J. E. Galan. 2001. A *Salmonella* inositol polyphosphatase acts in conjunction with other bacterial effectors to promote host cell actin cytoskeleton rearrangements and bacterial internalization. Mol Microbiol 39:248–260.
43. Zhou, D., M. S. Mooseker, and J. E. Galan. 1999. An invasion-associated *Salmonella* protein modulates the actin-bundling activity of plastin. Proc Natl Acad Sci USA 96:10176–81.
44. Zhou, D., M. S. Mooseker, and J. E. Galan. 1999. Role of the *S. typhimurium* actin-binding protein SipA in bacterial internalization. Science 283:2092–5.

CTMI (2005) 291:43–60

Regulation of Phagocytosis by Rho GTPases

F. Niedergang* · P. Chavrier (✉)

Membrane and Cytoskeleton Dynamics Group, Institut Curie, CNRS UMR144, 75248 Paris, France
Philippe.Chavrier@curie.fr

*Present Address: Department of Cell biology, Institut Cochin, CNRS UMR8104, INSERM U567, 75014 Paris, France

Abstract Phagocytosis is the mechanism of internalization used by specialized cells such as macrophages, dendritic cells, and neutrophils to internalize, degrade, and eventually present peptides derived from particulate antigens. The phagocytic process comprises several sequential and complex events initiated by the recognition of ligands on the surface of the particles by specific receptors on the surface of the phagocytic cells. Receptor clustering at the attachment site generates a phagocytic signal that in turn leads to local polymerization of actin filaments and to particle internalization. Depending on the particles and receptors involved, it appears that the structures and mechanisms associated with particle ingestion are diverse. However, work during the past few years has highlighted the importance of small GTP-binding proteins of the Rho family in various types of phagocytosis. As reviewed here, Rho family GTPases, their activators, and their downstream effectors control the local reorganization of the actin cytoskeleton beneath bound particles.

Abbreviations

Arp2/3	Actin-related protein 2/3
CR	Complement receptor
DH	Dbl-homology
FcγR	Fcγ receptor
GAP	GTPase-activating protein
GEF	Guanine nucleotide exchange factor
GDP	Guanosine 5′-diphosphate
GTP	Guanosine 5′-triphosphate
ITAM	Immuno-receptor tyrosine-based activation motif
PAK1	p21-Activated kinase 1
PI3K	Phosphatidylinositol 3′-kinase
PIP_2	Phosphatidylinositol 4,5-bisphosphate
PIP_3	Phosphatidylinositol 3,4,5-trisphosphate
SH2	Src homology 2
ROK	Rho-kinase
VASP	Vasodilator-stimulated phosphoprotein
WASP	Wiskott-Aldrich syndrome protein

1 Introduction

Phagocytosis is a universal cell function, which couples the recognition and binding of a particle (over 0.5 μm in diameter), generally in a receptor-dependent manner, to its internalization and degradation [3, 92]. Single-cell eukaryotes such as the mold *Dictyostelium discoideum* and amoebae use phagocytosis for feeding. In higher organisms, phagocytosis is fundamental for host defense against invading pathogens and contributes to the immune and inflammatory responses [2, 37]. Phagocytosis is also important during development for normal turnover and remodeling of tissues and disposal of dead cells [77]. In mammals, phagocytosis is the hallmark of specialized cells including macrophages, dendritic cells, and polymorphonuclear neutrophils. These cells are collectively referred to as professional phagocytes [73]. In certain circumstances, other cell types, such as thyroid and bladder epithelial cells or mesangial cells in the kidney, are able to perform phagocytosis. Receptors on the surface of the phagocytes can be classified into two main classes: receptors for opsonins such as IgG antibodies and the complement fragment C3bi that engage FcγRs and complement receptors (CR), respectively, and nonopsonic receptors. The latter can be subdivided into two groups, the Toll-like receptors (TLRs) and the non-Toll-like receptors, including scavenger receptors, C-type lectins, and C-type lectin-like receptors, which recognize components on the particle surface such as mannose or fucose residues, phosphatidylserine, and lipopolysaccharides [3, 89].

Actin polymerization beneath the site of attachment of the particle is the driving force behind ingestion and proceeds from signal transduction downstream of the phagocytic receptors (for review, see [23, 35, 64]). Even though the precise signaling cascades linking activated receptors to actin polymerization are not fully understood as yet, it has become clear that Rho GTPases control the cytoskeletal rearrangements during uptake of opsonized particles and apoptotic bodies by professional mammalian phagocytes [18, 27, 55, 61]. Like all members of the Ras superfamily, Rho proteins cycle between an inactive conformation when bound to guanosine 5′-diphosphate (GDP), and an active, guanosine 5′-triphosphate (GTP)-bound state. Cycling between these two states is regulated by guanine nucleotide exchange factors (GEFs), which promote GDP dissociation and GTP binding, and GTPase activating proteins (GAPs), which stimulate the low intrinsic GTPase activity of Rho proteins [90]. In the GTP-bound state, Rho proteins interact with downstream effectors to control actin filament assembly and organization into complex structures involved in cell shape, motility, and polarity [34]. In particular, at the leading edge of motile cells, Cdc42 and Rac1 regulate actin polymerization to form filopodia and lamellipodia, respectively, whereas in the cell body RhoA induces assembly of focal adhesions and contractile actin-myosin stress fibers. In this chapter, we will focus on recent advances made in the understanding of the regulation of actin dynamics during phagocytosis by Rho GTPases. Phagocytosis mediated by receptors for opsonins, and especially FcRs, has been the subject of most studies. We will first describe the molecular networks involving Rho GTPases in FcγR-mediated phagocytosis and then focus on the other phagocytic pathways.

2
Control of FcγR-Mediated Phagocytosis by Rac1/Cdc42 in Macrophages

FcγRI and FcγRIIIa are associated with low-molecular-weight γ-subunit homodimers, which contain an immunoreceptor tyrosine-based activation motif (ITAM) in their cytoplasmic region (Fig. 1) [75]. On receptor clustering by IgG-opsonized particles, the tyrosines in the ITAM are phosphorylated by Src family kinases and serve as docking sites for Src homology 2 (SH2) domain-containing cytosolic proteins. Among these, the tyrosine kinase Syk and the

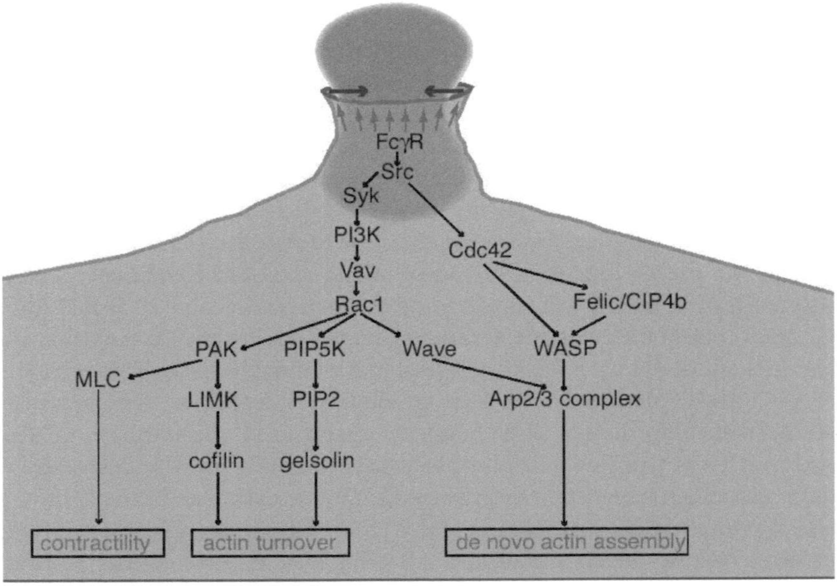

Fig. 1 Signaling in FcR-mediated phagocytosis. Rac1 and Cdc42 are the Rho GTPases controlling this pathway. IgG-coated particles bind to FcR and induce their aggregation, which activates Src family tyrosine kinases that, in turn, phosphorylate and activate the Syk kinase. Cdc42, activated by an unidentified GEF, recruits WASP, which stimulates actin nucleation by the Arp2/3 complex. Felic/CIP4b is another Cdc42-effector that may contribute to activate WASP. Rac1 is activated by Vav, which GEF activity is modulated by tyrosine phosphorylation and PI3K products (PIP_3). Downstream of Rac1, the Wawe/Scar complex may contribute to stimulate actin nucleation. Rac1-dependent activation of PIP5 K leads to the accumulation of PIP_2 that contributes to actin reorganization. PAK, another Rac1 effector activated during FcR-mediated phagocytosis, may play a role in actin turnover by activating the LIMK, which in turn phosphorylates and inhibits cofilin. PAK also phosphorylates myosin chains and therefore controls phagosomal contractility

p85 subunit of type I phosphatidylinositol 3′-kinase (PI3K) are essential for phagocytosis as they trigger the tyrosine phosphorylation of multiple proteins (Syk) and the local accumulation of phosphatidylinositol 3,4,5-trisphosphate (PIP_3) (PI3K) (for review see [14, 23, 35, 64]). Actin assembly at the site of particle attachment is induced within seconds on FcR ligation, giving rise to a ringlike structure, the actin cup, which surrounds the particle and rapidly disassembles as the particle becomes enclosed in the phagosome [36, 42]. Experiments using bacterial inhibitory toxins (see chapter by Baldwin and Barbieri, this volume) or expression of dominant inhibitory mutant forms to impair the activity of Rho GTPases demonstrated that actin organization in the phagocytic cup is controlled by Cdc42 and Rac, whereas RhoA is not involved [18, 27, 61]. Rac1 and Cdc42 are recruited to the site of particle attachment and accumulate transiently as GTP-loaded forms [18, 66, 70]. Dynamic studies by fluorescence resonance energy transfer (FRET) revealed different patterns of activation for Rac and Cdc42. GTP-Cdc42 was found at the tip of the advancing pseudopod, where it colocalized with polymerizing actin, whereas Rac1 activation was biphasic. GTP-Rac1 was induced at a low level early after particle binding and peaked at the time of pseudopod fusion [46]. Rac2, a close homolog of Rac1 expressed exclusively in hematopoietic cells, was also activated at the time of pseudopod fusion. The primary function of Rac2 on the phagosomal membrane seems to be regulation of superoxide production by the NADPH oxidase complex ([46]; see chapter by Bokoch, this volume). These data together with earlier findings [20, 61], suggested that Cdc42 and Rac1 regulate distinct processes during phagocytosis, with Cdc42 controlling pseudopod extension and Rac1 potentially regulating phagosome closure.

2.1
Mechanism of Rac/Cdc42 Activation During FcR-Mediated Phagocytosis

The cascade of events leading to Rac/Cdc42 activation downstream of FcRs is not fully understood. Nucleotide exchange on Rho GTPases is catalyzed by GEFs belonging to a family comprising at least 50 members in humans, which contain a Dbl-homology (DH) catalytic domain flanked by a pleckstrin homology (PH) domain [53]. The Vav family of Rho GEFs comprises three members (Vav1, -2, -3), which share conserved DH-PH and carboxy-terminal SH2-SH3-SH2 modules but have distinct tissue expression patterns (for review, see [15]). Inhibition of Vav activity prevents both Rac1 activation and FcγR-mediated phagocytosis in macrophages [70]. Nucleotide exchange activity of Vav proteins is modulated by tyrosine phosphorylation, and Syk-mediated phosphorylation of Vav on FcγR ligation is probably instrumental for Rac activation during phagocytosis [15]. The accumulation of PIP_3 in the phagocytic cup may also contribute to Vav activation, as phosphatidylinositol lipids are known to influence the enzymatic activity of Vav proteins through binding to their PH domain [41, 60]. Interestingly, inhibition of PI3K by wortmannin (wtn), although it abolishes phagocytosis, does not impair initial actin assembly beneath the particle, suggesting that the main function of a cascade linking PI3K, Vav, and Rac1 may be to control pseudopod extension and fusion [7, 28]. So far, the GEF(s) responsible for Cdc42 activation during FcR-mediated phagocytosis remains unidentified.

2.2 Regulation of Actin Dynamics Downstream of Rac/Cdc42 During FcR-Mediated Phagocytosis

In their GTP-bound active conformation, Rac1 and Cdc42 interact with downstream effectors that promote actin filament assembly and shape the filaments to form a phagocytic cup. One key effector is the Wiskott-Aldrich syndrome (WAS) protein (WASP), a protein expressed by hematopoietic cells that is recruited to the phagocytic cup [19, 25]. The lack of functional WASP protein in macrophages from WAS patients results in a deficient uptake of IgG-opsonized particles, indicating its central role during FcR-mediated phagocytosis (for review see [85]). Work from many laboratories has shown that WASP exists in an autoinhibited conformation that can be relieved through direct interaction with GTP-Cdc42 and phosphatidylinositol 4,5-bisphosphate (PIP_2) acting synergistically. Therefore, together with local Cdc42 activation, the transient accumulation of PIP_2 in the nascent phagocytic cup could be part of an activation signal for WASP ([11]; also see below). Another potential activator of WASP that is recruited to the phagocytic cup is a protein called Felic/CIP4b [31]. Felic/CIP4b, a partner for both GTP-bound Cdc42 and WASP [9, 86], shares extensive similarities with a protein called Toca-1 recently shown to contribute to Cdc42-dependent activation of WASP [44]. In turn, WASP stimulates the actin-nucleating activity of the Arp2/3 complex [72]. The Arp2/3 complex, which contains two actin-related proteins, is a major regulator of actin assembly that stimulates de novo actin polymerization by filament branching on existing filaments [72]. Therefore, WASP-mediated activation of the Arp2/3 complex in the phagocytic cup probably accounts for filament assembly that drives pseudopod extension downstream of Cdc42 [63]. In addition, the nucleating activity of the Arp/3 complex can be stimulated by the WAVE/Scar proteins, which are part of a multiprotein complex acting downstream of Rac1 [32, 65]. The finding that components of the WAVE-Scar complex are required for phagocytosis in *Drosophila* cells suggests that GTP-Rac1 may provide additional signals to Arp2/3 complex activation in the phagocytic cup during FcR-mediated phagocytosis [71, 74].

Additional activities are likely to participate in remodeling of the actin cup. A complex consisting of WASP, the adaptor proteins Fyb/SLAP, Nck, and SLP-76, and the cytoskeletal Ena/vasodilator-stimulated phosphoprotein (VASP) assembles in response to FcγR clustering and localizes to the cup [25]. VASP may also form a cooperative complex with WASP at the plasma membrane [19]. Ena/VASP proteins were shown to reduce the density of Arp2/3-dependent actin filament branches by increasing the rate of dissociation of filaments from the branches [79]. They were also reported to associate with fast-growing barbed ends of actin filaments, thereby antagonizing binding of capping proteins [50]. All together, these findings suggest an important contribution of Ena/VASP to the dynamics of actin filaments within the cup.

The number and availability of barbed ends are affected by proteins that sever the filaments and others that cap them. Gelsolin severs or caps actin filaments depending on Ca^{2+} and PIP_2 levels, thus providing a controlled way of generating free barbed ends (for review, see [78]). Neutrophils from gelsolin-deficient mice show defects in FcR-mediated phagocytosis that is impaired in both particle attachment and ingestion, demonstrating a role for gelsolin at an early step during phagocytosis [81]. Gelsolin is downstream of Rac1, as GTP-bound Rac1 stimulates actin-gelsolin dissociation [8, 10]. Uncapping of actin filaments may be promoted by Rac-dependent activation of phosphatidylinositol-4-phosphate 5-kinase (PIP5K) and PIP_2 production, as PIP_2 triggers the dissociation of the actin-gelsolin complex [87, 93]. PIP_2 accumulates in the phagocytic cup within regions of F-actin enrichment [11], and one isoform of PIP5K (PIP5KIα) is recruited at the cup, where its activity is required for actin reorganization and particle uptake [24]. Besides Rho GTPases, ADP-ribosylation factor 6 (ARF6), which belongs to the ARF family of small GTP-binding proteins (for review, see [22]), is also known to regulate the activity of PIP5K [45]. ARF6 is activated on FcR clustering with kinetics similar to that of Rac1/Cdc42 and is required for the ingestion of IgG-coated particles [66, 96]. Therefore, multiple pathways may converge to ensure that de novo synthesis of PIP_2 occurs at the nascent cup.

Last but not least, cofilin has been shown to play a role in phagocytosis of serum-opsonized zymosan (presumably involving several types of phagocytic receptors) in monocytic and macrophage cell lines [1]. Cofilin severs and depolymerizes F-actin filaments, thereby providing G-actin subunits to support the rapid-turnover barbed-end growth of filaments driving the extension of the plasma membrane (for review, see [16, 69]). Translocation of cofilin to the phagocytic cup occurs in activated neutrophils and macrophages and is associated with a cycle of phosphorylation/dephosphorylation [1, 43]. Cofilin activity is blocked by phosphorylation of a serine residue at position 3, and kinase cascades linking Cdc42 and Rac1 (and RhoA, see below) to cofilin have been described. The Ser/Thr p21-activated kinase 1 (PAK1), which is a downstream effector of Rac1 and Cdc42 that accumulates on phagosomal membrane [29], stimulates LIM kinase (LIMK) by phosphorylation on residue 508 [33]. In turn, activated LIMK directly phosphorylates cofilin at position 3, thereby blocking its activity [95]. Therefore, the PAK/LIMK/cofilin pathway could mediate Rac/Cdc42 control of actin filament turnover during phagocytosis of opsonized particles [62].

2.3 Regulation of Phagosomal Contractility

Contractile activities are necessary for phagocytosis, and they most likely involve interaction between actin filaments and molecular motors of the myosin superfamily [21]. When adjacent macrophages attempting to ingest a single erythrocyte were examined, dumbbell-shaped erythrocytes were observed with their two bulbous ends still connected by a thin membrane stalk [84]. In the presence of Wtn/LY294002 or butanedione monoxime (BDM, an inhibitor of myosins), constricted erythrocytes were absent, arguing for a role for PI3K and myosins in the generation of contractility [84]. Several myosin motors (I, II, V, IX, and X), which are recruited to the phagosome at different stages of its formation, could conceivably control the generation of force during phagocytosis [6, 30, 68, 84]. Indeed, myosin X that is recruited to the phagocytic cup through binding of PI3K products by its PH domains has been identified as a key downstream effector of PI3K required for optimal extension of pseudopods during FcR phagocytosis [26]. Several laboratories have also shown that myosin II is required for FcR-mediated phagocytosis [6, 59, 68]. Interestingly, Rac/Cdc42 could potentially regulate myosin II contractility through alteration of the phosphorylation status of myosin light and heavy chains, possibly mediated by PAK [29, 80, 91].

3 Phagocytosis Mediated by Other Receptors

3.1 Complement Receptor-Mediated Phagocytosis

Complement activation via the alternative pathway leads to the deposition of complement fragment C3bi on the surface of the particle that is recognized by the phagocytic C3bi receptor (CR3), corresponding to the integrin $\alpha_M\beta_2$ (also called CD11b/CD18 or Mac-1) (Fig. 2) [3]. In contrast to FcRs that are constitutively active for phagocytosis, activation of CR3 requires extrastimuli such as chemokines, TNF-α, or adhesion to fibronectin-coated surfaces or can be experimentally induced by phorbol esters. Recently, FcR ligation has also been shown to promote clustering of CR3 into high-avidity complexes capable of binding C3bi-coated targets, revealing possible cross talk between phagocytic receptors [47]. Activation of CR3 is therefore based on an inside-out signaling that appears to be controlled by the small GTP-binding protein Rap1 (for review see [17]). Only limited pointlike attachment sites are established between C3bi-coated particles and the phagocyte sur-

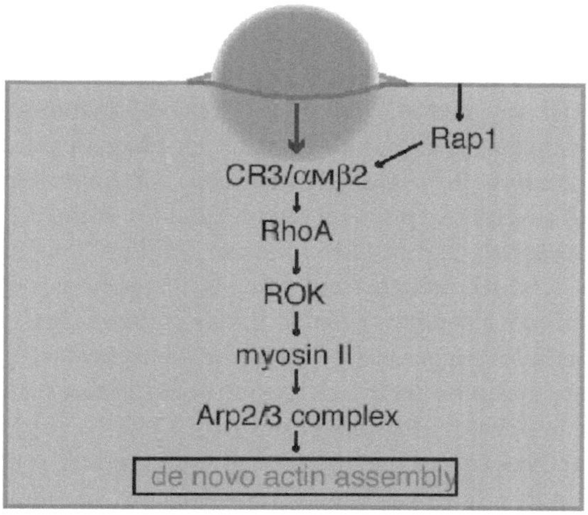

Fig. 2 Signaling in CR-mediated phagocytosis. Phagocytosis is triggered by interaction between C3bi-coated particles and the CR3/$\alpha M\beta 2$ receptor. This integrin receptor has to be activated by an "inside-out" signal implicating the GTP-binding protein Rap1 in order to acquire an active conformation. RhoA activated downstream of CR3/$\alpha M\beta 2$ in turn recruits the Rho kinase and its target myosin-II that have been involved in the accumulation of the Arp2/3 complex and actin nucleation

face, and importantly, during ingestion, the particle sinks directly into the cytoplasm of the phagocyte without the extension of membrane pseudopods that are typical of FcR-mediated phagocytosis [48]. By immunofluorescence, these contact sites appear as foci enriched in F-actin and various cytoskeletal proteins including paxillin, vinculin, α-actinin, and Arp2/3 complex [5, 63].

The function of Rho GTP-binding proteins in CR3-mediated phagocytosis has been examined by Caron and colleagues, who reported that inhibition of RhoA by *C. botulinum* C3 exotoxin blocks CR3-mediated uptake and interferes with Arp2/3 accumulation at the contact site [18, 63]. In contrast, inhibition of Cdc42 or Rac1 does not affect CR3-mediated phagocytosis, providing a molecular basis for the distinct morphological features of FcR- and CR3-mediated phagocytosis [18]. The RhoA effector Rho-kinase (ROK) and its target myosin II have been implicated in the accumulation of Arp2/3 complex and F-actin assembly during CR3-mediated phagocytosis; however, their precise mechanism of action remains unknown [68]. ROK, which directly phosphorylates LIMK [67, 83], can also regulate cofilin phosphorylation with biological consequences during Rho-mediated neurite retraction [58]. Similarly, the ROK/LIMK pathway could mediate RhoA control of actin filament turnover during phagocytosis downstream of CR3.

3.2
Phagocytosis of Zymosan

Zymosan is a cell wall preparation of heat-treated *Saccharomyces cerevisiae* that has been widely used as a model particle for phagocytosis. The preparation is mainly composed of β-glucans, mannans, mannoproteins, and chitins. It interacts with several receptors on the phagocytes' surface, including CR3, the mannose receptors that bind to mannan residues, and the recently described Dectin-1, which recognizes β-glucans. Phagocytosis of unopsonized zymosan is therefore a complex event that may engage several receptors, and the signaling cascades triggered are likely to be multiple as well. Dectin-1 is an ITAM-containing receptor likely to trigger tyrosine kinases and downstream signals similar to FcRs. It also cooperates with TLR2 receptors that connect to the NF-κB activation cascade [52]. Small GTP-binding proteins have not been involved in these pathways so far. Interestingly, Rac and Cdc42 appear to be important for the uptake of nonopsonized zymosan via CR3 receptors, whereas serum-opsonized zymosan, which binds to CR3 as well, leads to a pathway depending on Rho as described for other complement-opsonized targets [51]. Therefore, depending on the target, the CR3 receptor may be engaged via distinct epitopes and trigger different signaling cascades relying on different Rho GTPases.

3.3
Uptake of Apoptotic Targets

Cells undergoing the process of apoptosis are taken up by professional phagocytes as well as by nonprofessional neighboring cells, and in this case engulfment is usually not followed by activation of proinflammatory responses but is rather a "silent" event. As for other types of phagocytosis, entry relies on actin polymerization. The receptors involved in uptake of apoptotic targets are numerous, including the $\alpha_V\beta_5$- and $\alpha_V\beta_3$-integrin receptors, a phosphatidylserine receptor, and scavenger receptors such as CD36 and CD68 (Fig. 3) (for review, see [38]). Studies in *C. elegans* have unraveled two pathways important for engulfment of apoptotic cells, implicating ced-2, ced-5, and ced-10 on one hand and ced-1, ced6, and ced-7 on the other hand. Recent work in mammalian cells revealed that the signaling cascades existing in the worm are evolutionarily conserved. Interestingly, the ortholog of CED10 is Rac1, and DOCK180/CED5 is an unconventional GEF for Rac1. DOCK180/CED5 does not contain a DH domain and bears instead, as a domain responsible for the exchange activity, a domain called Docker. The adaptor CrkII/CED2 is believed to recruit DOCK180/CED5 and ELMO/CED12 in a ternary complex

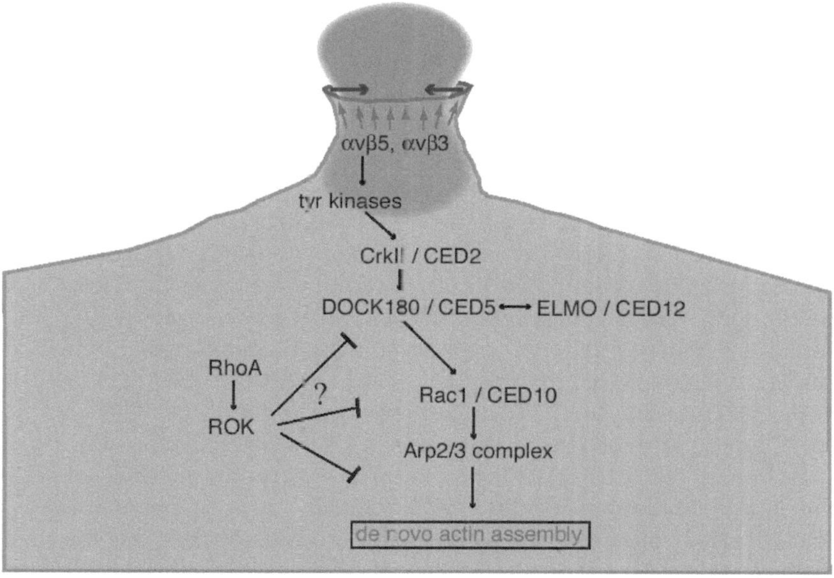

Fig. 3 Signaling during engulfment of apoptotic cells. Internalization of apoptotic cells is induced by the recognition of many receptors. Studies in mammals and *C. elegans* have demonstrated the pathways triggered by the $\alpha_V\beta_3$, $\alpha_V\beta_5$ integrins and the CED1, CED7 receptors. Interaction of apoptotic targets with the $\alpha_V\beta_3$ and $\alpha_V\beta_5$ integrins stimulate tyrosine kinases. The adaptor CrkII/CED2 recruits a complex composed of DOCK180/CED5 and ELMO/CED12 that is an exchange factor for Rac1/CED10. The Arp2/3 complex is stimulated and participates in actin polymerization. Basal engulfment activity is inhibited by RhoA and its effector ROK, by a mechanism that is so far not described, and inhibition by RhoA can be counterbalanced by the stimulation of the CrkII/CED2, DOCK180/CED5, ELMO/CED12 complex

that is able to exchange GTP on Rac1/CED10 [4, 13, 39, 76]. ELMO/CED12 indeed contains a PH domain that may contribute to the subcellular localization of the complex and the control of GEF activity [57]. Therefore, the DOCK180/CED5-ELMO/CED12 complex may be considered as the equivalent of a Dbl-family GEF, containing both a DH and a PH domain in the same molecule [12]. Overexpression of DOCK180/CED5 and ELMO/CED12 increased the efficiency of phagocytosis of beads or apoptotic cells. Because there are five mammalian DOCK180 family members and three ELMO isoforms, there may be finely tuned interplays between all these molecules, depending on the context and the cell function. In addition, another Rho GTPase, RhoG, binds to the N-terminal part of ELMO2 in its GTP-bound form [49]. RhoG is known to play a role in neurite outgrowth and cell spreading, by activating Rac on NGF or integrin stimulation, respectively, and this interaction places the DOCK180/CED5-ELMO/CED12 complex as a target of RhoG that activates Rac1 in these signaling pathways [49]. Whether RhoG also plays a role in integrin-mediated uptake of phagocytic cells is not yet known.

Beside the positive regulation of engulfment by Rac1, a negative regulation of this pathway by Rho has been recently described in mammalian cells [55, 88]. Initial observations indicated that RhoA was downregulated during engulfment and that inhibition of RhoA by C3 toxin enhanced phagocytosis of apoptotic targets or carboxylate beads. In addition, expression of active RhoA or Rho-GEFs reduced phagocytosis efficiency. Therefore, it appears that Rho negatively regulates the basal engulfment activity and that this inhibition can be counterbalanced by activation of the CrkII/CED2, DOCK180/CED5, ELMO/CED12, and Rac1/CED10 pathway. One of the RhoA effectors, the ROK, may be preferentially implicated in this negative regulation [88].

Activation of Rac1/CED10 on binding of an apoptotic target then connects the signaling cascade with the Arp2/3 complex and actin polymerization. WASP is also implicated in uptake of apoptotic targets, because it is recruited in phagocytic cups and because engulfment is delayed in macrophages derived from WASP-deficient mice [54].

The ced1, ced6, and ced7 products have also mammalian equivalents. CED1/LRP (low-density lipoprotein receptor-related protein) is related to the mammalian SREC (scavenger receptor for endothelial cells), whereas CED7 and its mammalian counterpart ABC1 are ATP-binding cassette (ABC) transporters. They are both involved in surface recognition of apoptotic targets [40, 94, 97]. CED6 is a downstream effector, because its overexpression bypasses the requirement for ced-1 and ced-7 and because it partially rescues *ced-10/Rac1* mutants [56]. The mammalian counterpart of CED6 is a protein called GULP for enGULfment adaPter Protein [82]. It acts as an adaptor protein and binds to the cytoplasmic tail of CED1/LRP [82]. Downstream effectors are still to be identified as well as the potential links between the ced-2, ced-5, ced-10 and the ced-1, ced6, ced-7 pathways.

4 Conclusion

Although entry into cells via phagocytosis can take many forms, in most cases, regulation of actin dynamics is under the control of Rho family GTPases. Interestingly, in *C. elegans* as well as in mammalian cells, the molecular mechanisms underlying the internalization of large particles are often similar to the mechanisms controlling cell migration. With the development of large-scale screens for effectors of the Rho GTPases, there has been much progress recently in our understanding of the networks downstream of the Rho proteins. However, many "black boxes" remain upstream in the signaling cascades and in the way signals are integrated from the many receptors that can be stimulated on the surface of a phagocytic cell. Microbial pathogens that mimic or exploit many of the signaling pathways induced by the stimulation of phagocytic receptors may help us unravel some of the remaining questions.

References

1. Adachi R, Takeuchi K, Suzuki K (2002) Antisense oligonucleotide to cofilin enhances respiratory burst and phagocytosis in opsonized zymosan-stimulated mouse macrophage J774.1 cells. J Biol Chem 277:45566–45571
2. Aderem A (2003) Phagocytosis and the inflammatory response. J Infect Dis 187 Suppl 2:S340–S345
3. Aderem A, Underhill DM (1999) Mechanisms of phagocytosis in macrophages. Annu Rev Immunol 17:593–623
4. Albert ML, Kim JI, Birge RB (2000) $\alpha_V\beta_5$ Integrin recruits the CrkII-Dock180-rac1 complex for phagocytosis of apoptotic cells. Nat Cell Biol 2:899–905
5. Allen LA, Aderem A (1996) Molecular definition of distinct cytoskeletal structures involved in complement- and Fc receptor-mediated phagocytosis in macrophages. J Exp Med 184:627–637
6. Araki N, Hatae T, Furukawa A, Swanson JA (2003) Phosphoinositide-3-kinase-independent contractile activities associated with Fcγ-receptor-mediated phagocytosis and macropinocytosis in macrophages. J Cell Sci 116:247–257
7. Araki N, Johnson MT, Swanson JA (1996) A role for phosphoinositide 3-kinase in the completion of macropinocytosis and phagocytosis by macrophages. J Cell Biol 135:1249–1260
8. Arcaro A (1998) The small GTP-binding protein Rac promotes the dissociation of gelsolin from actin filaments in neutrophils. J Biol Chem 273:805–813
9. Aspenstrom P (1997) A Cdc42 target protein with homology to the non-kinase domain of FER has a potential role in regulating the actin cytoskeleton. Curr Biol 7:479–487
10. Azuma T, Witke W, Stossel TP, Hartwig JH, Kwiatkowski DJ (1998) Gelsolin is a downstream effector of rac for fibroblast motility EMBO J 17:1362–1370
11. Botelho RJ, Teruel M, Dierckman R, Anderson R, Wells A, York JD, Meyer T, Grinstein S (2000) Localized biphasic changes in phosphatidylinositol-4,5-bisphosphate at sites of phagocytosis. J Cell Biol 151:1353–1368
12. Braga VM (2002) GEF without a Dbl domain? Nat Cell Biol 4:E188–E190.
13. Brugnera E, Haney L, Grimsley C, Lu M, Walk SF, Tosello-Trampont AC, Macara IG, Madhani H, Fink GR, Ravichandran KS (2002) Unconventional Rac-GEF activity is mediated through the Dock180-ELMO complex. Nat Cell Biol 4:574–582.
14. Brumell JH, Grinstein S (2003) Role of lipid-mediated signal transduction in bacterial internalization. Cell Microbiol 5:287–297
15. Bustelo XR (2000) Regulatory and signaling properties of the Vav family. Mol Cell Biol 20:1461–1477

16. Carlier MF, Ressad F, Pantaloni D (1999) Control of actin dynamics in cell motility. Role of ADF/cofilin. J Biol Chem 274:33827–33830
17. Caron E (2003) Cellular functions of the Rap1 GTP-binding protein: a pattern emerges. J Cell Sci 116:435–440
18. Caron E, Hall A (1998) Identification of two distinct mechanisms of phagocytosis controlled by different rho GTPases. Science 282:1717–1721
19. Castellano F, Le Clainche C, Patin D, Carlier MF, Chavrier P (2001) A WASp-VASP complex regulates actin polymerization at the plasma membrane. EMBO J 20:5603–5614.
20. Castellano F, Montcourrier P, Chavrier P (2000) Membrane recruitment of rac1 triggers phagocytosis. J Cell Sci 113:2955–2961
21. Chavrier P (2002) May the force be with you: Myosin-X in phagocytosis. Nat Cell Biol 4:E169–R171
22. Chavrier P, Goud B (1999) The role of ARF and rab GTPases in membrane transport. Curr Opin Cell Biol 11:466–475
23. Chimini G, Chavrier P (2000) Function of rho family proteins in actin dynamics during phagocytosis and engulfment. Nat Cell Biol 2:E191–E196
24. Coppolino MG, Dierckman R, Loijens J, Collins RF, Pouladi M, Jongstra-Bilen J, Schreiber AD, Trimble WS, Anderson R, Grinstein S (2002) Inhibition of phosphatidylinositol-4-phosphate 5-kinase Iα impairs localized actin remodeling and suppresses phagocytosis. J Biol Chem 277:43849–43857
25. Coppolino MG, Krause M, Hagendorff P, Monner DA, Trimble W, Grinstein S, Wehland J, Sechi AS (2001) Evidence for a molecular complex consisting of Fyb/SLAP, SLP-76, Nck, VASP and WASP that links the actin cytoskeleton to Fcγ receptor signalling during phagocytosis. J Cell Sci 114:4307–4318
26. Cox D, Berg JS, Crammer M, O'Chinegwundoh JO, Dale BM, Cheney RE, Greenberg S (2002) Myosin-X as a downstream effector of PI 3-kinase during phagocytosis. Nat Cell Biol 4:
27. Cox D, Chang P, Zhang Q, Reddy PG, Bokoch GM, Greenberg S (1997) Requirements for both rac1 and cdc42 in membrane ruffling and phagocytosis in leukocytes. J Exp Med 186:1487–1494
28. Cox D, Tseng CC, Bjekic G, Greenberg S (1999) A requirement for phosphatidylinositol 3-kinase in pseudopod extension. J Biol Chem 274:1240–1247
29. Dharmawardhane S, Brownson D, Lennartz M, Bokoch GM (1999) Localization of p21-activated kinase 1 (PAK1) to pseudopodia, membrane ruffles, and phagocytic cups in activated human neutrophils. J Leukoc Biol 66:521–527
30. Diakonova M, Bokoch G, Swanson JA (2002) Dynamics of cytoskeletal proteins during Fcγ receptor-mediated phagocytosis in macrophages. Mol Biol Cell 13:402–411.
31. Dombrosky-Ferlan P, Grishin A, Botelho RJ, Sampson M, Wang L, Rudert WA, Grinstein S, Corey SJ (2003) Felic (CIP4b), a novel binding partner with the Src kinase Lyn and Cdc42, localizes to the phagocytic cup. Blood 101:2804–2809
32. Eden S, Rohatgi R, Podtelejnikov AV, Mann M, Kirschner MW (2002) Mechanism of regulation of WAVE1-induced actin nucleation by Rac1 and Nck. Nature 418:790–793
33. Edwards DC, Sanders LC, Bokoch GM, Gill GN (1999) Activation of LIM-kinase by Pak1 couples Rac/Cdc42 GTPase signalling to actin cytoskeletal dynamics. Nat Cell Biol 1:253–259
34. Etienne-Manneville S, Hall A (2002) Rho GTPases in cell biology. Nature 420:629–635

35. Greenberg S (1995) Signal transduction of phagocytosis. Trends Biochem Sci 5:93–97
36. Greenberg S, el Khoury J, di Virgilio F, Kaplan EM, Silverstein SC (1991) Ca^{2+}-independent F-actin assembly and disassembly during Fc receptor-mediated phagocytosis in mouse macrophages. J.Cell Biol 113:757–767
37. Greenberg S, Grinstein S (2002) Phagocytosis and innate immunity. Curr Opin Immunol 14:136–145.
38. Grimsley C, Ravichandran KS (2003) Cues for apoptotic cell engulfment: eat-me, don't eat-me and come-get-me signals. Trends Cell Biol 13:648–656
39. Gumienny TL, Brugnera E, Tosello-Trampont AC, Kinchen JM, Haney LB, Nishiwaki K, Walk SF, Nemergut ME, Macara IG, Francis R, Schedl T, Qin Y, Van Aelst L, Hengartner MO, Ravichandran KS (2001) CED-12/ELMO, a novel member of the CrkII/Dock180/Rac pathway, is required for phagocytosis and cell migration. Cell 107:27–41.
40. Hamon Y, Broccardo C, Chambenoit O, Luciani MF, Toti F, Chaslin S, Freyssinet JM, Devaux PF, McNeish J, Marguet D, Chimini G (2000) ABC1 promotes engulfment of apoptotic cells and transbilayer redistribution of phosphatidylserine. Nat Cell Biol 2:399–406
41. Han J, Luby-Phelps K, Das B, Shu X, Xia Y, Mosteller RD, Krishna UM, Falck JR, White MA, Broek D (1998) Role of substrates and products of PI 3-kinase in regulating activation of rac-related guanosine triphosphatases by Vav. Science 279:558–560
42. Henry RM, Hoppe AD, Joshi N, Swanson JA (2004) The uniformity of phagosome maturation in macrophages. J Cell Biol 164:185–194
43. Heyworth PG, Robinson JM, Ding J, Ellis BA, Badwey JA (1997) Cofilin undergoes rapid dephosphorylation in stimulated neutrophils and translocates to ruffled membranes enriched in products of the NADPH oxidase complex. Evidence for a novel cycle of phosphorylation and dephosphorylation. Histochem Cell Biol 108:221–233
44. Ho HY, Rohatgi R, Lebensohn AM, Le M, Li J, Gygi SP, Kirschner MW (2004) Toca-1 mediates Cdc42-dependent actin nucleation by activating the N-WASP-WIP complex. Cell 118:203–216
45. Honda A, Nogami M, Yokozeki T, Yamazaki M, Nakamura H, Watanabe H, Kawamoto K, Nakayama K, Morris AJ, Frohman MA, Kanaho Y (1999) Phosphatidylinositol 4-phosphate 5-kinase $\alpha$is a downstream effector of the small G protein ARF6 in membrane ruffle formation. Cell 99:521–532
46. Hoppe AD, Swanson JA (2004) Cdc42, Rac1, and Rac2 Display distinct patterns of activation during phagocytosis. Mol Biol Cell 15:3509–3519
47. Jongstra-Bilen J, Harrison R, Grinstein S (2003) Fcγ-receptors induce Mac-1 (CD11b/CD18) mobilization and accumulation in the phagocytic cup for optimal phagocytosis. J Biol Chem 278:45720–45729
48. Kaplan G (1977) Differences in the mode of phagocytosis with Fc and C3 receptors in macrophages. Scand J Immunol 6:797–807
49. Katoh H, Negishi M (2003) RhoG activates Rac1 by direct interaction with the Dock180-binding protein Elmo. Nature 424:461–464
50. Krause M, Dent EW, Bear JE, Loureiro JJ, Gertler FB (2003) Ena/VASP proteins: regulators of the actin cytoskeleton and cell migration. Annu Rev Cell Dev Biol 19:541–564
51. Le Cabec V, Carreno S, Moisand A, Bordier C, Maridonneau-Parini I (2002) Complement receptor 3 (CD11b/CD18) mediates type I and type II phagocytosis during nonopsonic and opsonic phagocytosis, respectively. J Immunol 169:2003–2009

52. Lemaitre B (2004) The road to Toll. Nat Rev Immunol 4:521–527
53. Letunic I, Copley RR, Schmidt S, Ciccarelli FD, Doerks T, Schultz J, Ponting CP, Bork P (2004) SMART 4.0: towards genomic data integration. Nucleic Acids Res 32 Database issue:D142–144
54. Leverrier Y, Lorenzi R, Blundell MP, Brickell P, Kinnon C, Ridley AJ, Thrasher AJ (2001) The Wiskott-Aldrich syndrome protein is required for efficient phagocytosis of apoptotic cells. J Immunol 166:4831–4834
55. Leverrier Y, Ridley AJ (2001) Requirement for Rho GTPases and PI 3-kinases during apoptotic cell phagocytosis by macrophages. Curr Biol 11:195–199
56. Liu QA, Hengartner MO (1998) Candidate adaptor protein CED-6 promotes the engulfment of apoptotic cells in *C. elegans*. Cell 93:961–972
57. Lu M, Kinchen JM, Rossman KL, Grimsley C, deBakker C, Brugnera E, Tosello-Trampont AC, Haney LB, Klingele D, Sondek J, Hengartner MO, Ravichandran KS (2004) PH domain of ELMO functions in trans to regulate Rac activation via Dock180. Nat Struct Mol Biol 11:756–762
58. Maekawa M, Ishizaki T, Boku S, Watanabe N, Fujita A, Iwamatsu A, Obinata T, Ohashi K, Mizuno K, Narumiya S (1999) Signaling from rho to the actin cytoskeleton through protein kinases ROCK and LIM-kinase. Science 285:895–898
59. Mansfield PJ, Shayman JA, Boxer LA (2000) Regulation of polymorphonuclear leukocyte phagocytosis by myosin light chain kinase after activation of mitogen-activated protein kinase. Blood 95:2407–2412.
60. Marshall JG, Booth JW, Stambolic V, Mak T, Balla T, Schreiber AD, Meyer T, Grinstein S (2001) Restricted accumulation of phosphatidylinositol 3-kinase products in a plasmalemmal subdomain during Fcγ receptor-mediated phagocytosis. J Cell Biol 153:1369–1380.
61. Massol P, Montcourrier P, Guillemot JC, Chavrier P (1998) Fc receptor-mediated phagocytosis requires CDC42 and Rac1. EMBO J 17:6219–6229
62. Matsui S, Matsumoto S, Adachi R, Kusui K, Hirayama A, Watanabe H, Ohashi K, Mizuno K, Yamaguchi T, Kasahara T, Suzuki K (2002) LIM Kinase 1 modulates opsonized zymosan-triggered activation of macrophage-like U937 Cells. possible involvement of phosphorylation of cofilin and reorganization of actin cytoskeleton. J Biol Chem 277:544–549
63. May RC, Caron E, Hall A, Machesky LM (2000) Involvement of the Arp2/3 complex in phagocytosis mediated by FcgR or CR3. Nat Cell Biol
64. May RC, Machesky LM (2001) Phagocytosis and the actin cytoskeleton. J Cell Sci 114:1061–1077
65. Miki H, Suetsugu S, Takenawa T (1998) WAVE, a novel WASP-family protein involved in actin reorganization induced by Rac. EMBO J. 17:6932–6941
66. Niedergang F, Colucci-Guyon E, Dubois T, Raposo G, Chavrier P (2003) ADP ribosylation factor 6 is activated and controls membrane delivery during phagocytosis in macrophages. J Cell Biol 161:1143–1150
67. Ohashi K, Nagata K, Maekawa M, Ishizaki T, Narumiya S, Mizuno K (2000) Rho-associated kinase ROCK activates LIM-kinase 1 by phosphorylation at threonine 508 within the activation loop. J Biol Chem. 275:3577–3582
68. Olazabal IM, Caron E, May RC, Schilling K, Knecht DA, Machesky LM (2002) Rho-kinase and myosin-II control phagocytic cup formation during CR, but not FcgammaR, phagocytosis. Curr Biol 12:1413–1418
69. Pantaloni D, Le Clainche C, Carlier MF (2001) Mechanism of actin-based motility. Science 292:1502–1506
70. Patel JC, Hall A, Caron E (2002) Vav regulates activation of Rac but not Cdc42 during FcgγR-mediated phagocytosis. Mol Biol Cell 13:1215–1226.

71. Pearson AM, Baksa K, Ramet M, Protas M, McKee M, Brown D, Ezekowitz RA (2003) Identification of cytoskeletal regulatory proteins required for efficient phagocytosis in *Drosophila*. Microbes Infect 5:815–824
72. Pollard TD, Borisy GG (2003) Cellular motility driven by assembly and disassembly of actin filaments. Cell 112:453–465
73. Rabinovitch M (1995) Professional and non-professional phagocytes: an introduction. Trends Cell Biol 5:85–87
74. Ramet M, Manfruelli P, Pearson A, Mathey-Prevot B, Ezekowitz RA (2002) Functional genomic analysis of phagocytosis and identification of a *Drosophila* receptor for *E. coli*. Nature 416:644–648
75. Ravetch JV, Bolland S (2001) IgG Fc receptors. Annu Rev Immunol 19:275–290
76. Reddien PW, Horvitz HR (2000) CED-2/CrkII and CED-10/Rac control phagocytosis and cell migration in *Caenorhabditis elegans*. Nat Cell Biol 2:131–136
77. Ren Y, Savill J (1998) Apoptosis: the importance of being eaten. Cell Death Differ 5:563–568
78. Revenu C, Athman R, Robine S, Louvard D (2004) The co-workers of actin filaments: from cell structures to signals. Nat Rev Mol Cell Biol 5:635–646
79. Samarin S, Romero S, Kocks C, Didry D, Pantaloni D, Carlier MF (2003) How VASP enhances actin-based motility. J Cell Biol 163:131–142
80. Sells MA, Boyd JT, Chernoff J (1999) p21-Activated kinase 1 (Pak1) regulates cell motility in mammalian fibroblasts. J.Cell Biol 145:837–849
81. Serrander L, Skarman P, Rasmussen B, Witke W, Lew DP, Krause K-H, Stendahl O, Nubetae O (2000) Selective inhibition of IgG-mediated phagocytosis in gelsolin-deficient murine neutrophils. J Immunol 165:2451–2457
82. Su HP, Nakada-Tsukui K, Tosello-Trampont AC, Li Y, Bu G, Henson PM, Ravichandran KS (2002) Interaction of CED-6/GULP, an adapter protein involved in engulfment of apoptotic cells with CED-1 and CD91/low density lipoprotein receptor-related protein (LRP). J Biol Chem 277:11772–11779
83. Sumi T, Matsumoto K, Nakamura T (2001) Specific activation of LIM kinase 2 via phosphorylation of threonine 505 by ROCK, a Rho-dependent protein kinase. J.Biol Chem 276:670–676
84. Swanson JA, Johnson MT, Beningo K, Post P, Mooseker M, Araki N (1999) A contractile activity that closes phagosomes in macrophages. J Cell Sci 112:307–316
85. Thrasher AJ (2002) WASp in immune-system organization and function. Nat Rev Immunol 2:635–646
86. Tian L, Nelson DL, Stewart DM (2000) Cdc42-interacting protein 4 mediates binding of the Wiskott-Aldrich syndrome protein to microtubules. J Biol Chem 275:7854–7861
87. Tolias KF, Hartwig JH, Ishihara H, Shibasaki Y, Cantley LC, Carpenter CL (2000) Type Iα phosphatidylinositol-4-phosphate 5-kinase mediates Rac-dependent actin assembly. Curr Biol 10:153–156
88. Tosello-Trampont AC, Nakada-Tsukui K, Ravichandran KS (2003) Engulfment of apoptotic cells is negatively regulated by Rho-mediated signaling. J Biol Chem 278:49911–49919
89. Underhill DM, Ozinsky A (2002) Phagocytosis of microbes: complexity in action. Annu Rev Immunol 20:825–852.
90. Van Aelst L, D'Souza-Schorey C (1997) Rho GTPases and signaling networks. Genes Dev 11:2295–2322
91. van Leeuwen FN, van Delft S, Kain HE, van der Kammen RA, Collard JG (1999) Rac regulates phosphorylation of the myosin-II heavy chain, actinomyosin disassembly and cell spreading. Nat Cell Biol 1:242–248

92. Vieira OV, Botelho RJ, Grinstein S (2002) Phagosome maturation: aging gracefully. Biochem J 366:689–704
93. Weernink PA, Meletiadis K, Hommeltenberg S, Hinz M, Ishihara H, Schmidt M, Jakobs KH (2004) Activation of type I phosphatidylinositol 4-phosphate 5-kinase isoforms by the Rho GTPases, RhoA, Rac1, and Cdc42. J Biol Chem 279:7840–7849
94. Wu YC, Horvitz HR (1998) The *C. elegans* cell corpse engulfment gene ced-7 encodes a protein similar to ABC transporters. Cell 93:951–960
95. Yang N, Higuchi O, Ohashi K, Nagata K, Wada A, Kangawa K, Nishida E, Mizuno K (1998) Cofilin phosphorylation by LIM-kinase 1 and its role in Rac-mediated actin reorganization. Nature 393:809–812
96. Zhang Q, Cox D, Tseng CC, Donaldson JG, Greenberg S (1998) A requirement for ARF6 in fcγ receptor-mediated phagocytosis in macrophages. J Biol Chem 273:19977–19981
97. Zhou Z, Hartwieg E, Horvitz HR (2001) CED-1 is a transmembrane receptor that mediates cell corpse engulfment in *C. elegans*. Cell 104:43–56

CTMI (2005) 291:61–90

The Immunological Synapse and Rho GTPases

M. Deckert (✉) · C. Moon · S. Le Bras

INSERM Unit 576, Hôpital de l'Archet, BP3079, 06202 Nice Cedex 3, France
deckert@unice.fr

Abstract Rho GTPases are molecular switches controlling a broad range of cellular processes including lymphocyte activation. Not surprisingly, Rho GTPases are now recognized as pivotal regulators of antigen-specific T cell activation by APCs and immunological synapse formation. This review summarizes recent advances in our understanding of how Rho GTPase-dependent pathways control T lymphocyte motility, polarization and activation.

Abbreviations

TCR	T cell receptor
MHC	Major histocompatibility complexes
APC	Antigen-presenting cell
DC	Dendritic cell
CTL	Cytotoxic T lymphocyte

NK	Natural killer
MTOC	Microtubule organizing center
SMAC	Supramolecular activation cluster
PTK	Protein tyrosine kinase
GEF	Guanosine nucleotide exchange factor
ERM	Ezrin radixin moesin proteins

1 Introduction

A major task of lymphocytes is to circulate continuously throughout the body, where they scan the surface of antigen-presenting cells (APC) for the presence of pathogen-derived peptide antigens bound to major histocompatibility complex (MHC) molecules. Whereas $CD8^+$ cytotoxic T cells kill cells that are replicating intracellular pathogens through secretion of lytic enzymes, activated $CD4^+$ T cells help mount an efficient immune response by secreting immunomodulatory cytokines. This recognition depends on the ability of the T cell antigen receptor (TCR) to specifically bind to the peptide-MHC complex, and just a few copies of an antigenic peptide displayed on the MHC surface molecules of the APC can trigger a robust T cell activation, a pivotal event in adaptative immunity. To succeed in this task, T lymphocytes engage a series of extremely dynamic and narrow contacts with APCs that take place in a submicrometer-scale gap between the two cells.

This interaction zone, recently referred to as the immunological synapse (IS), is exquisitely organized in time and space and characterized by polarization of T cells toward the APC, movements of T cell membrane receptors binding to their ligands expressed at the APC membrane, and recruitment and activation of signaling proteins. The organization of the IS, which depends on the nature of the T cell-APC pair, is largely regulated by the reorganization of T cell actin and tubulin cytoskeleton. Although first noted more than 20 years ago, the implication of cytoskeletal structures during lymphocyte activation has only received recent experimental and molecular support. Rho GTPases are molecular switches controlling a broad range of cellular processes (for reviews see Etienne-Manneville and Hall 2002 and accompanying reviews), including lymphocyte activation (Altman and Deckert 1999; Cantrell 2003). Not surprisingly, Rho GTPases are now recognized as pivotal regulators of antigen-specific T cell activation by APCs and IS formation. This review summarizes recent advances in our understanding of how Rho GTPase-dependent pathways control T lymphocyte motility, activation, and effector functions.

2
Immunological Synapses: More than One Face

T cell activation requires recognition by the TCR of MHC-peptide complexes present at the surface of APCs. This interaction induces different functional programs depending on the nature and the maturation stage of the T lymphocyte. In thymocytes, binding of MHC-peptide to TCR will either induce positive selection, resulting in maturation of T cells capable of recognizing foreign peptides in the context of MHCs, or negative selection, resulting in apoptosis of autoreactive thymocytes. In $CD8^+$ cytotoxic T lymphocytes (CTL), recognition of the MHC-peptide complexes on the target cell induces killing of the target. In $CD4^+$ helper T lymphocytes, binding of the TCR to specific MHC-peptide complexes at the surface of B lymphocytes induces secretion by T cells of cytokines that allow the interacting B cells to differentiate in antibodies producing cells. Finally, TCR recognition of MHC-peptide complexes present on dendritic cells (DCs) allows naive T cells that have never encountered the antigen for which they are specific to differentiate, secrete cytokines, and proliferate. The term immunological synapse was proposed more than 20 years ago to describe the interaction observed between T cells and APCs, which is characterized by a stable, narrow contact zone between the two cells. Subsequent multidimensional fluorescence analysis of the cell–cell interface revealed that T cell recognition of MHC-peptide complexes at the surface of targets or B cells was accompanied by a dynamic spatial organization of membrane receptors, cytoskeleton, and intracellular signaling complexes on the T cell side (Kupfer and Kupfer 2003; Kupfer and Singer 1989). The IS prototype was first described by Kupfer and colleagues as the organized structure formed between T cell clones and antigen-bearing B lymphoma cells, the so-called bull's-eye pattern (Monks et al. 1998). After 15 min of contact, T cell membrane receptors and signaling molecules redistribute into concentric supramolecular activation clusters (SMACs) at the T cell–APC interface. During the maturation process of the IS and the formation of the bull's-eye pattern, a central SMAC (c-SMAC) is formed where TCRs accumulate and from where integrins such as LFA-1 and their ligands are gradually excluded to form a peripheral SMAC (p-SMAC) (Monks et al. 1998) (Fig. 1). High-speed microscopy studies with fluorescent chimera of TCRζ chain further demonstrated that the IS is highly dynamic (Krummel et al. 2000).

Other membrane molecules such as the large adhesion receptor CD43 (Allenspach et al. 2001; Delon et al. 2001) and the transmembrane tyrosine phosphatase CD45 (Johnson et al. 2000) were found to be excluded from the contact zone. These observations were confirmed in other studies using

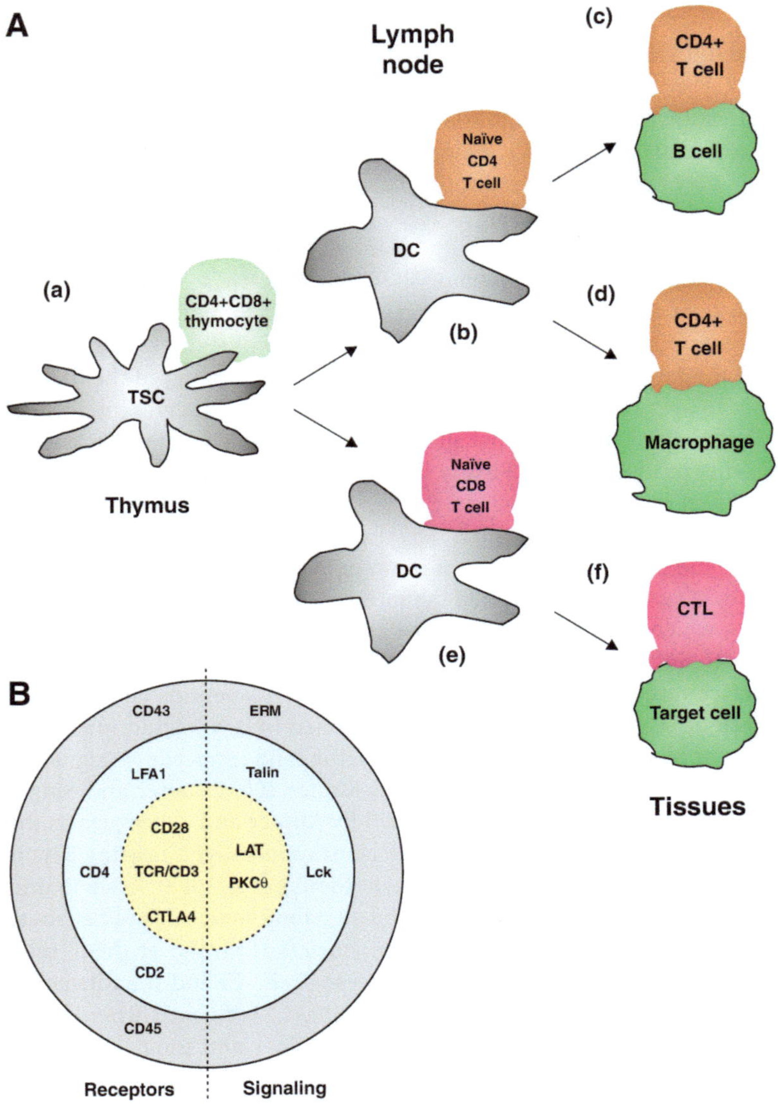

Fig. 1A, B. Immunological synapses. **A** Different types of Immunological synapses. Interaction between thymocytes and thymic stromal cell (*TSC*) (*a*); naive CD4$^+$ T cell and dendritic cell (*DC*) (*b*); effector CD4$^+$ T cell and B cell (*c*); effector CD4$^+$ T cell and macrophage (*d*); naive CD8$^+$ T cell and DC (*e*); effector CD8$^+$ cytotoxic T cell (*CTL*) and target cell (*f*). **B** Typical concentric organization of a mature synapse in a CD4$^+$ T cell interacting with an agonist-loaded B cell ("bull's-eye" pattern), including the c-SMAC, the p-SMAC, and a more distal region of exclusion, as well as the localization of the major cell surface receptors (*left*) and signaling molecules (*right*)

artificial planar bilayers containing fluorescent peptide-MHC molecules and ICAM-1 (Grakoui et al. 1999) and Jurkat leukemic T cells interacting with a B lymphoma pulsed with a superantigen (Blanchard et al. 2002a). Patterning of membrane receptors is also accompanied by a redistribution of signaling molecules. Indeed, Lck, ZAP-70 (Lee et al. 2002), PKCθ (Blanchard et al. 2002a; Monks et al. 1997), and LAT (Blanchard et al. 2002a; Monks et al. 1997) localize at the center of the T cell–APC interface after TCR activation, whereas the actin linker talin (Monks et al. 1998) and the ERM family protein moesin (Delon et al. 2001) are found in the peripheral zone.

Since then, several IS showing highly different organization patterns have been recognized according to the T–APC pair of cells involved (reviewed in Trautmann and Valitutti 2003). This includes the interaction between CD4^{+} helper T cells and B lymphocytes, CD8^{+} cytotoxic T cells and target cells, T cells and DCs, and between thymocytes and thymic stromal cells (Fig. 1). IS between natural killer (NK) cells and their targets have also been described (Davis 2002). Of note, the concentric pattern is not consistently observed, for example, when peptides are presented to naive T cells by DCs (Revy et al. 2001), during lymphocyte crawling on the surface of a DC (Friedl and Brocker 2002) or during thymic selection of immature CD4^{+} CD8^{+} thymocytes (Hailman et al. 2002), suggesting that the formation of a bull's-eye pattern is not absolutely required for T cell response. However, recent studies using biphotonic confocal microscopy further indicated that effective IS are also produced in intact lymphoid organs when T cells encounter DCs in lymph nodes (Cahalan et al. 2002) or when thymocytes interact with thymic stromal cells (Bousso et al. 2002). Thus more than one pattern has been observed, and the term IS can cover distinct cell–cell conjugates and systems. This term is now largely accepted by investigators for any stable and flattened interface between a lymphocyte or NK cell and a recognized cell, associated with receptors and cytoskeletal reorganization (Huppa and Davis 2003). As stated above, a common feature of the multiple described IS is a high degree of temporal and spatial molecular organization, with discrete patterns of surface and intracellular signaling molecules at the interface. Although patterning of the molecules at the cell–cell interface may occur spontaneously, as proposed by mathematical modeling taking into account membrane fluidity, protein size, and receptor—ligand affinity (Huppa and Davis 2003; Shaw and Dustin 1997), several studies have unraveled the critical role of active processes such as Rho GTPase-dependent cytoskeletal movement in driving receptor clustering and synapse formation (Acuto and Cantrell 2000; Valitutti et al. 1995; Wulfing and Davis 1998).

3
Rho GTPases in Lymphocyte Circulation and Motility

T cells are mobile cells trafficking throughout the body to scan for exogenous antigens. During the development of immune responses and before interacting with APCs, activated T cells respond to chemokine signals that attract them from the blood into target inflamed tissue in a process called extravasation (reviewed in Barreiro et al. 2004). This process is also important during the constant recirculation of unchallenged T cells throughout the lymphoid organs (Campbell et al. 1998, 2003). Extravasation involves tethering and rolling of lymphocytes on the blood vessel wall, firm adhesion, and diapedesis through the endothelial barrier. Activated endothelium exhibits several adhesion molecules, which orchestrate the different steps of lymphocyte migration, including chemokines, selectins, and integrins ligands such as VCAM1 and ICAM1, which respectively bind β_1- and β_2-integrins found on the surface of circulating lymphocytes (Ardouin et al. 2003; Friedl and Brocker 2002). Lymphocyte tethering to endothelial cells is mediated by the interaction of selectins expressed by most leukocytes with their counterreceptors. A critical step in the control of lymphocyte rolling is provided by the interaction between L-selectin and sialylated molecules expressed on endothelial cells. Although no clear role of Rho GTPases during this process has been described so far, their implication is likely because L-selectin cytoplasmic tail interacts with moesin (Ivetic et al. 2002), a member of the ERM family of membrane-actin cytoskeleton linkers that participate in the actin-dependent formation of microvilli (or filopodia), required for rolling and tethering (McEver 2002; Yonemura and Tsukita 1999). Moreover, in lymphoblasts, P-selectin glycoprotein ligand 1 (PSGL-1), the best-characterized selectin ligand, also interacts with moesin, which recruits the protein tyrosine kinase (PTK) Syk (Urzainqui et al. 2002), a well-characterized upstream activator of cytoskeletal remodeling in lymphocytes (Altman and Deckert 1999; Dustin and Chan 2000). These findings suggest that both selectins and selectin ligands can signal to actin cytoskeleton-dependent structures during the early steps of lymphocyte tethering and rolling. Whether Rho GTPases directly participate in these processes remains, however, to be elucidated. After rolling and before extravasation, lymphocytes firmly arrest on the surface of the inflamed endothelium. This process is mediated through the binding of endothelial VCAM-1 to $\alpha_4\beta_1$ (VLA-4) and ICAM-1 to the β_2-integrin LFA1, found on the surface of circulating lymphocytes. Actin cytoskeleton reorganization affects integrin functions and cell adhesiveness through a distinct array of membrane-actin cytoskeleton linkers (Brakebusch and Fassler 2003; Hogg et al. 2003). Interestingly, actin cytoskeleton also actively participates in VCAM-1 and ICAM-1 clustering on endothelial cells through dynamic interaction with moesin (Barreiro et al. 2002). Conversely, increased actin polymerization and interaction with integrin tails can also affect integrin clustering and adhesive properties, characterizing a process referred to as "inside-out signaling." In this regard, TCR engagement increases integrin activity through several regulators of actin cytoskeleton, including PTKs of the Syk family and VAV guanine nucleotide exchange factors (GEFs) (Maneiro 2000). Another important regulator of this TCR-to-integrin signaling appears to be the adaptor protein ADAP (adhesion- and degranulation-promoting adaptor protein, also named SLAP-130 or Fyb) (Krause et al. 2000). ADAP$^{-/-}$ mice display defective T cell activation and proliferation, which result from impaired TCR-induced integrin-mediated adhesion (Griffiths et al. 2001; Peterson et al. 2001) (see below).

Lymphocyte adhesiveness directly influences the random motility of lymphocytes on the substratum, a process called haptotaxis. Smith et al. found that LFA-1-induced T cell random migration on ICAM-1 involves a coordinated regulation of MLCK-mediated attachment and ROCK-dependent detachment that requires a spatial segregation of kinases activity. MLCK and its activator calmodulin operated at the F-actin-enriched T cell leading edge, whereas ROCK and RhoA controlled the detachment of the T cell trailing edge (Smith et al. 2003). Rac1 may also be involved in T cell morphology and motility, because the engagement of the integrin LFA-1 in T cells has been shown to induce a transient activation of Rac1 through Vav1 and PI3K/Akt, which modulates T cell elongation on ICAM-1 (Sanchez-Martin et al. 2004).

The next step of transmigration is not well understood except that it clearly involves LFA-1/ICAM-1 interaction and contacts between lymphocyte integrins and junctional adhesion molecules (JAMs), which normally act as a molecular glue to maintain the integrity of the endothelial barrier (reviewed in Luscinskas et al., 2002). However, migrating lymphocytes crawl across the endothelial cell–cell junction in a process involving GTPase-dependent cytoskeletal reorganization on both cell sides and massive cell shape modifications (del Pozo et al. 1999; Vicente-Manzanares et al. 2002). Thus the sequential activation of spatially organized Rho GTPase signaling pathways allows the integration of the molecular cues issued from chemotactic and adhesive receptors involved in lymphocyte motility, polarization, and migration.

4
Rho GTPases in Lymphocyte Polarization Toward the APC

After extravasation, lymphocytes must move toward APCs localized within the target tissues. Note that target tissues can be different depending on the T cell type. For example, whereas naive T cells will enter lymphoid organs where antigens are presented by DCs, memory T cells will migrate into inflammatory peripheral tissues to be activated by antigens displayed by macrophages. The directional movement of lymphocytes results from a combination of substratum-dependent motility (see above section) and directed migration regulated by extracellular signals such as chemotactic gradients. The best-characterized chemotactic receptors on lymphocytes are the chemokine receptors, a subset of the G protein-coupled receptor (GPCR) family. Chemokines are small proteins that can either be soluble or immobilized on extracellular matrix or cells. In addition to governing cell migration, chemokines regulate actin polymerization and morphological changes, adhesion through modulation of integrin functions, gene transcription, and survival (Moser and Loetscher 2001). Under a chemotactic gradient, the motile lymphocyte becomes highly polarized and displays a characteristic morphology, with the nucleus pushed into the leading edge and the cytoplasm concentrated in the rear extension of the cell called the uropod (McFarland 1969). This cellular polarization is also accompanied by an asymmetric distribution of receptors and signaling molecules. Integrins (Sanchez-Madrid and del Pozo 1999), TCRs coreceptors (Krummel et al. 2000), and large molecules such as CD43, CD44, and ICAMs are found in the uropod (del Pozo et al. 1995), together with cytoplasmic organelles including endoplasmic reticulum and Golgi apparatus, microtubule organizing center (MTOC), and secretory vesicles (Kupfer et al. 1987; Vicente-Manzanares and Sanchez-Madrid 2004). On the contrary, chemotactic receptors are found in the leading edge, where they are thought to promote the formation of filopodia extensions involved in T cell orientation (Nieto et al. 1997). Initial studies with dominant interfering Cdc42 mutants expressed in T cells showed that Cdc42 regulates T cell polarization toward the APC (Stowers et al. 1995), suggesting that this process is controlled by chemotactic receptors localized at the T cell leading edge. In experiments using T cells from WASP-deficient patients and inhibitory CRIB fusion proteins, the interaction between Cdc42 and WASP has been further implicated during T cell chemotaxis induced by the chemokine SDF-1/CXCL12, a ligand for the CXCR4 chemokine receptor (Haddad et al. 2001). Interestingly, we and others have shown that the cytoplasmic PTK ZAP-70, a known activator of Rac and Cdc42 GTPases in T cells, was required for SDF-1/CXCL12-induced T cell chemotaxis and migration

(Ottoson et al. 2001; Ticchioni et al. 2002). It remains to be determined, however, how APCs control these directed movements and also the nature of the T cell specific GEFs and Rho GTPases effectors involved during this process.

Within the first minute of contact with an APC, T cells adhere transiently to the APC and scan it for the presence of the appropriate MHC-peptide complexes. The first adhesion step involves interaction between the integrin LFA-1 and their ligands ICAM-1 and ICAM-3 (Montoya et al. 2002) and can occur in the absence of antigen as shown for naive T cell–DC synapse (Delon et al. 1998; Revy et al. 2001). Initial scanning involves morphological changes and is accompanied by a low level of calcium pulses (Donnadieu et al. 1994) that may favor long-term survival of the naive T cell (Revy et al. 2001) and may facilitate antigen recognition. When found, the specific MHC-peptide combination triggers TCR-mediated signaling, which first promotes stable T cell–APC interactions. This event, called the "stop signal" (Dustin et al. 1997), must compete with chemokine signals to prevent the T cell from further chemotaxis (Bromley et al. 2000). The stop signal relies on rapid (within seconds) biochemical signals triggered by TCR (Beeson et al. 1996) and is characterized by the release of intracellular calcium whose magnitude is influenced by the peptide identity (Bachmann et al. 1997; Wulfing et al. 1997). It is not known, however, whether Rho GTPases are involved during this process.

5 TCR Signaling at the T Cell–APC Interface: Shaping up the IS

The morphological changes occurring during lymphocyte polarization and activation were first described than 20 years ago (Haston et al. 1982), and the first molecular indication that Rho GTPases could regulate T cell activation was provided by Lang et al. (Lang et al. 1992). More recent work with inhibitors of actin polymerization and myosin motors has further shown that the remodeling of actin cytoskeleton induced by TCR engagement is necessary to receptor patterning and IS formation (Wulfing and Davis 1998). Strong evidence that Rho GTPases are involved in T cell activation later came from genetic studies. In mice lacking Rac2, a hematopoietic specific isoform of Rac proteins, actin reorganization in T cells is impaired (Yu et al. 2001).

Studies using transgenic animals have shown critical roles for Rac1 (Gomez et al. 2000, 2001) and Cdc42 (Na et al. 1999) in T cell development and activation. Furthermore, transgenic expression in lymphocytes of the bacterial C3 toxin that ADP-ribosylates and inactivates RhoA revealed the major function of RhoA in thymic development and mature T cell activation (Corre et al.

2001; reviewed in Cantrell 2003). However, the most compelling evidence for the implication of Rho GTPases during T cell activation and IS formation was provided by studies on regulators and effectors of Rho GTPase pathways. TCR engagement at the T cell–APC interface activates the Src family kinases Lck or Fyn, which phosphorylate tyrosine residues in the immunoreceptor tyrosine-based activation motifs (ITAMs) of the cytoplasmic tails of the CD3–TCR complex. Phosphorylated ITAMs then bind the Syk family kinase

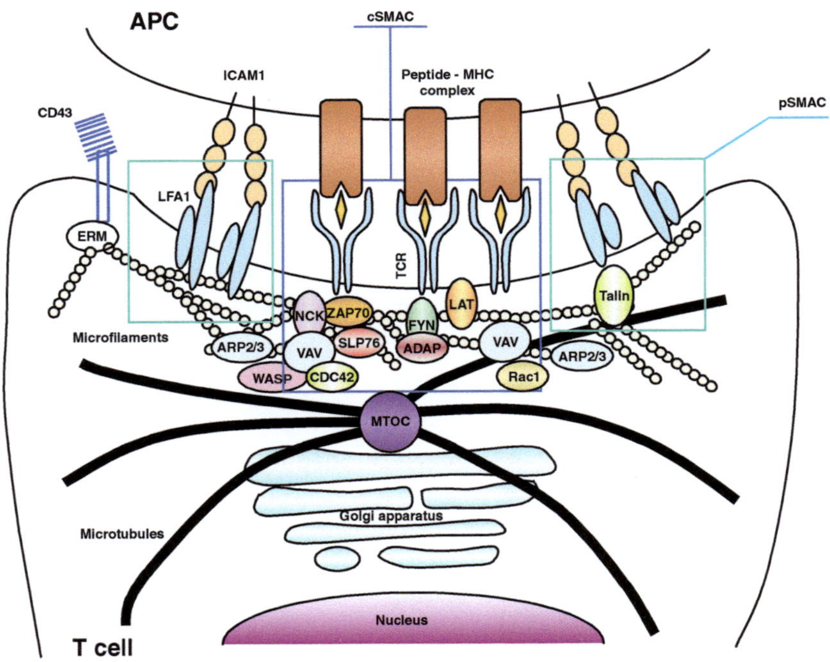

Fig. 2 Model of Rho GTPase-dependent cytoskeleton reorganization at the Immunological synapse. TCR engagement by peptide-MHC complexes on APC initiates a signaling cascade involving the activation of protein tyrosine kinases such as Fyn and ZAP-70, the recruitment of adaptor proteins such as LAT, SLP-76, Nck, and ADAP, and the activation of actin polymerization-regulatory proteins such as the guanine nucleotide exchange factor Vav1, the Rho GTPases Cdc42 and Rac1, WASP, and Arp2/3 complex. The resulting localized actin polymerization regulates the formation of supramolecular activation clusters (*SMACs*) and the reorientation of the microtubule organizing center (*MTOC*) and the Golgi apparatus beneath the T cell–APC contact zone. Actin microfilament remodeling is also regulated through LFA1 interaction with the cytolinker talin and through CD43 interaction with ezrin-radixin-moesin (*ERM*) proteins found outside the SMACs

ZAP-70 through its tandem SH2 domains (Chan et al. 1994). Recruitment of ZAP-70 leads to its phosphorylation and enzymatic activation, allowing the phosphorylation of enzymes and adaptor proteins, thereby activating several signaling pathways, including phospholipase C-γ (PLC-γ), small GTPases of Ras and Rho families, lipid kinases, and serine/threonine kinases such as PKCs and MAPKs (Altman and Deckert 1999; Samelson 2002). Among the downstream effectors of TCR signaling, the adaptor proteins LAT and SLP-76 (Jackman et al. 1995; Tomlinson et al. 2000; Zhang et al, 1998), and the VAV family of Rac/Cdc42 GEFs (Collins et al. 1997; Deckert et al. 1996; Turner and Billadeau 2002) are undoubtedly central masters of lymphocyte signaling and cytoskeleton remodeling (Fig. 2).

The Vav family is composed of three conserved genes, Vav1, Vav2, and Vav3. Whereas Vav1 expression is restricted to hematopoietic cells, Vav2 and Vav3 display a much broader tissue expression (Turner and Billadeau 2002). Vav GEFs are highly homologous proteins composed of a catalytic Dbl-homology (DH) domain—the hallmark of all known Rho-GEFs—and other structural domains involved in protein-protein or protein-lipid interaction and signal transduction. Vav proteins represent important integrators of Rho GTPase signaling pathways downstream of PTK-coupled receptors including immunoreceptors (TCR, BCR, and FcR) and integrins. The catalytic activity of Vav GEFs is activated by tyrosine phosphorylation (Crespo et al. 1997), and Vav1 SH2 domain has been shown to bind several PTKs including Syk family kinases (Collins et al. 1997; Deckert et al. 1996). Furthermore, the accessibility of Vav1 DH domain for Rho substrates is regulated by a conformational change of its amino-terminal autoinhibitory domain involving the phosphorylation of a conserved tyrosine residue found in its acidic domain (Aghazadeh et al. 2000). The exact catalytic specificity of Vav GEFs in vivo remains unknown. However, their functional importance for lymphocyte development and effector functions has been demonstrated by genetic (Doody et al. 2001; Fujikawa et al. 2003; Tedford et al. 2001) and biochemical (Charvet et al. 2002; Doody et al. 2000; Tartare-Deckert et al. 2001; Zakaria et al. 2004) studies. Furthermore, studies using Vav1-deficient mice showed activation defects in T cells that resulted from impaired actin polymerization (Fischer et al. 1998; Holsinger et al. 1998) and PLC-γ1 regulation (Reynolds et al. 2002). In addition, Vav1 appears to be important during actin cytoskeleton-dependent integrin activation (Ardouin et al. 2003; Krawczyk et al. 2002). In studies by Ardouin et al., Vav1-deficient thymocytes showed a decreased ability to form conjugates with APCs, resulting from impaired TCR-induced activation of the integrin LFA-1. However, the characteristic IS patterning of proteins was apparently normal, whereas MTOC relocalization was defective in the absence of Vav1 (Ardouin et al. 2003). This phenotype is reminiscent of that found with ZAP-70 deficient T cells (Blanchard et al. 2002a), suggesting that Vav1 and its upstream activator ZAP-70 transduce signals to only a subset of cytoskeleton-dependent events at the IS. Another possible link between actin cytoskeleton and the establishment of the IS comes from the observation that actin cytoskeleton regulators modulate the clustering of lipid rafts, important specialized membrane microdomains involved in IS assembly and T cell signaling (Bi and Altman 2001). This event was positively controlled by a Vav1/Rac1-dependent pathway (Villalba et al. 2001), and negatively regulated through ERM protein interaction with the complex of adaptor proteins EBP50 and CBP (Itoh et al. 2002). How these molecular events affect T cell activation and function are, however, unclear.

Effectors of Rho proteins also participate to T cell activation by APC. For example, the ZAP-70 substrate and adaptor protein SLP-76 forms with Nck and Vav1 a trimolecular complex that regulates actin polymerization through the activation of Rac and its effectors PAK1 (Bubeck Wardenburg et al. 1998) and WASP (Zeng et al. 2003). In addition, WASP localizes at the T cell–APC interface, where it activates the actin-related protein 2/3 (Arp2/3) complex involved in localized actin polymerization (Krause et al. 2000). Other molecules involved in TCR-stimulated actin polymerization are also found at the T cell–APC interface, including the adaptor proteins ADAP and EVL, a protein of the Ena/VASP family (Krause et al. 2000; Peterson 2003). The observation that $WAS^{P-/-}$ (Snapper et al. 1998), $WIP^{-/-}$ (Anton et al. 2002), $vav1^{-/-}$ (Ardouin et al. 2003; Fischer et al. 1998; Holsinger et al. 1998), and $ADAP^{-/-}$ (Griffiths et al. 2001; Peterson et al. 2001) T cells are hyporesponsive to TCR stimulation supports the notion that actin cytoskeleton remodeling modulates receptor distribution and activity at the T cell–APC interface. In further support of this idea, Faure et al. showed that a rapid inactivation of ERM proteins through a Vav1/Rac1-dependent pathway triggered by TCR engagement uncouples the cortical actin cytoskeleton from the plasma membrane, thus increasing cellular plasticity and leading to more efficient IS formation (Faure et al. 2004) (Fig. 2).

Besides receptor clustering, T cell–APC interaction also results in intracellular segregation of T cell signaling molecules such as PKCθ (Monks et al. 1998). Unlike other PKCs, only PKCθ translocates to the detergent-insoluble cellular fraction, which mostly represents the actin cytoskeleton (Villalba et al. 2000). An intriguing connection between PKCθ and the cytoskeleton exists, as PKCθ localization appears to be regulated by PI3K and Vav1 activities (Villalba et al. 2002). However, although dominant-negative PKCθ blocks Vav1-dependent signals, such as activation of JNK, IL-2 promoter, and NFAT reporter genes, it shows no effect on actin polymerization induced by Vav1 (Villalba et al., 2000), placing PKCθ downstream of Vav1-dependent growth signals but not downstream of the Vav1-dependent pathway leading to actin polymerization. A possible scenario is that TCR engagement stimulates Vav1-dependent actin remodeling responsible for TCR clustering and PKCθ translocation to the c-SMAC, probably through its interaction with the PDZ domain-containing protein CARD11/CARMA1 recently found localized within the IS (Gaide et al. 2002; Pomerantz et al. 2002; Wang et al. 2004).

Nevertheless, the above studies suggest that cortical actin cytoskeleton plays an active role in receptor patterning through specialized membrane domains, adaptor proteins, and activators of Rho GTPase signaling. Although recent studies have shown that TCR engagement leads to IS recruitment of SLAT/IBP, a novel Rac1/Cdc42 GEF involved in Th2 differentiation (Gupta et al. 2003; Tanaka et al. 2003), it remains, however, to determine whether other T cell specific GEFs also participate in IS formation and T cell activation.

6 Microtubules, Rho GTPases and the Immunological Synapse

Microtubule dynamics also participates in IS formation although the underlying mechanisms are less well characterized. Nevertheless, Rho GTPase signaling pathways appear to play a key role, as shown by blocking experiments using mutant Rho GTPases. The transfection of a dominant-negative form of Cdc42 into a T cell hybridoma blocked MTOC relocalization to the T cell–APC contact site (Stowers et al. 1995) (Fig. 2). More recent studies using video imaging showed that during target cell killing by CTL, the MTOC is drawn vectorially to the contact site by a microtubule sliding mechanism, and that microtubules anchor to the pSMAC defined by the dense clustering of LFA-1 (Kuhn and Poenie 2002). This process may favor redistribution of secretory compartments along microtubules, thus allowing directional killing (below).

The molecular mechanisms controlling the polarized redistribution of microtubules during IS formation are not completely understood. However, several clues have emerged. MTOC polarization during T cell–APC interaction is regulated by an ITAM-dependent process that requires the phosphorylation of at least one ITAM by the Src kinase Lck (Lowin-Kropf et al. 1998), the activity of ZAP-70 (Blanchard et al. 2002a; Kuhne et al. 2003) and Vav1 (Ardouin et al. 2003) (our unpublished results), and the expression of the adaptor proteins LAT and SLP-76 (Kuhne et al. 2003). Interestingly, impaired MTOC polarization in ZAP-70-deficient T cells did not affect central clustering of CD3 and CD2 as well as exclusion of CD45 and CD43 but did alter the recruitment of LAT and PKCθ to the cSMAC. Previous studies have shown that TCR stimulation induces the phosphorylation of tubulin in T lymphocytes

(Ley et al. 1994) and its association with Fyn, ZAP-70, and the Rac GEF Vav (Huby et al. 1995; Marie-Cardine et al. 1995). Microtubules are also required for activation of other Rho GEFs in nonlymphoid cells, suggesting the existence of cross talk between actin and tubulin networks. One notion emerging from these observations is that microtubules may act as a reservoir for actin and tubulin cytoskeleton regulatory proteins. On antigenic stimulation, these signaling components may be transported to the IS by the mean of motor proteins and/or organelles.

Conversely, forced deacetylation of microtubules induced by the histone deacetylase HDAC6 at the T cell–APC interface impairs antigen-specific MTOC polarization, receptor patterning, and IL-2 production through dynamic instability of microtubules (Serrador et al. 2004). Interestingly, integrin activation promotes microtubule stabilization at the leading edge of migrating fibroblasts through a pathway that involves the PTK FAK, Rho, and the Rho effector mDia (Palazzo et al. 2004). It would be interesting to determine whether a similar process also occurs during antigen-specific TCR and/or LFA-1 engagement and how this could connect to HDAC activation at the T cell–APC interface.

7 Rho GTPases and Synaptic Functions

Although immune synapses would certainly facilitate cell–cell communications, the exact purpose of assembling synapses is still subject to debate (Huppa and Davis 2003; Trautmann and Valitutti 2003). The observation that receptor patterning at the T cell–APC interface was also accompanied by the relocalization of critical signaling molecules such as Lck and PKCθ (Monks et al. 1998) initially suggested that IS formation modulates early TCR signaling. However, this hypothesis was recently ruled out by the observation that TCR-mediated tyrosine phosphorylation events occur before the formation of the IS (Lee et al. 2002). Other nonexclusive functions have been proposed, including sustained T cell activation, polarized delivery of molecules, and protein recycling/degradation. The implication of actin and tubulin cytoskeleton in these events is detailed in the chapter by Diebold and Bokoch, this volume.

7.1 Sustained T Cell Activation

The first evidence of the role of actin cytoskeleton during sustained T cell activation was provided by studies from Valitutti et al., who showed that disrupting actin reorganization impairs sustained signaling from TCR and T cell activation (Valitutti et al. 1995). Since then, other studies have suggested that IS formation may affect sustained T cell activation, in part through the regulation of late signaling events such as serine/threonine kinase activation (Matthews et al. 2000; Monks et al. 1997). The T cell-specific PKC isoform PKCθ (Baier et al. 1993) was one of the first identified proteins recruited to the T cell–APC interface (Monks et al. 1997).

Relocalization of PKCθ at the IS correlated with a long-lasting catalytic activation (more than 2 h). Importantly, several modes of partial T cell activation unable to cause PKCθ translocation also failed to cause T-cell proliferation (Monks et al. 1997), suggesting that long-lasting phosphorylation events play an important role in propagating activation signals required for T cell activation.

Another event that may modulate late T cell activation and survival is the activation of PI3-kinases (PI3Ks), a family of enzymes producing 3′-phosphoinositides that promote the activation of PH domain-containing proteins such as Akt/PKB (Cantrell 2002). Studies using a GFP-Akt PH domain fusion protein as a marker of PI3K activity showed that the metabolism of 3′-phosphoinositides occurred at the T cell membrane, inside and outside the T cell–APC interface (Costello et al. 2002; Harriague and Bismuth 2002). Remarkably, a sustained PI3K activation was necessary to T cell proliferation induced by antigenic stimulation (Costello et al. 2002), consistent with previous observations that naive T cells become committed to proliferation after 20 h of TCR stimulation (Iezzi et al. 1998) and that continuous TCR-dependent PI3K activation was required for full effector potential (Huppa et al. 2003). The connection between Rho GTPases and PI3K pathways is far from being understood. However, PI3K is considered in some studies as a Rac effector, and recently the activation of the PI3K/Akt pathway in immature mouse thymocytes has been shown to require the expression of the Rac GEF Vav1 (Reynolds et al. 2002). Interestingly, using peripheral resting T cells from Vav1-deficient mice, we have shown that Vav1 activity is required for TCR-dependent Akt activation and cell cycle entry (our unpublished data). Considering the critical role played by Rho GTPases in cell proliferation and survival (Coleman et al. 2004; Sahai and Marshall 2002), one function of actin dynamics may be to trigger the selective and long-lasting recruitment at the IS of enzymes required for late T cell activation events.

7.2 Polarized Delivery of Molecules

One key role of synapses is certainly the polarized secretion by T cells of cytokine toward the APC or cytolytic proteins toward the target cells. This directional delivery of molecules may ensure that the secreted proteins stay confined within the narrow space between the two interacting cells, therefore avoiding dilution and bystander effects. This process has been described in the directed secretion of cytokines at the contacting helper T cell/B cell membranes (Kupfer et al. 1991). However, one of the best characterized examples of polarized secretion is the killing process by $CD8^+$ T lymphocytes (CTLs). Cells infected by viruses are specifically killed by CTLs through the delivery of lytic granules within the extracellular contact zone between the two cells (Peters et al. 1991). Early evidence supporting a role for the actin cytoskeleton during the killing process was provided by the observation that cytoskeletal proteins such as talin localized at cell–cell contact zone (Kupfer and Singer 1989) and that introduction of recombinant *Clostridium botulinum* C3 toxin into CTLs inhibited their cytolytic function (Lang et al. 1992). More recent studies showed that, similar to helper $CD4^+$ lymphocytes, CTLs form IS with the target cell, and SMACs assembled when TCR molecules of CTLs are engaged by APCs exhibit organization similar to that observed with helper $CD4^+$ T cells (see previous sections and Potter et al. 2001; Stinchcombe et al. 2001b). However, this IS also exhibits a specific feature represented by a secretory domain formed at the interface between the cytotoxic cell and the target cell through which CTLs deliver vesicles containing lytic proteins such as perforin and granzymes that induce target cell apoptosis (Clark and Griffiths 2003; Peters et al. 1991). On CTL activation, uncharacterized signals emanating from the IS instruct the MTOC and Golgi complex to polarize toward the IS, allowing docking and fusion of the secretory lysosome to the secretory domain (Stinchcombe et al. 2001b). Thus deciphering the mechanisms regulating the assembly of the CTL IS will certainly help to understand the mechanisms of polarized secretion of lytic granules. Studies of rare human diseases led to the identification of some components of the lytic granule exocytic pathway (Clark and Griffiths 2003). For example, adaptor protein 3 (AP-3) deficiency in Hermansky-Pudlak syndrome (HPS) results in impaired lysosomal sorting, deficient movement of the lytic granules along microtubules, and docking within the secretory domain of the IS (Clark et al. 2003). The deficiency of the small GTPase Rab27a in CTLs from patients with Griscelli syndrome leads to reduced cytotoxicity and cytolytic granule exocytosis and immunodeficiency (Menasche et al. 2000). With mice models of Griscelli syndrome, it was further shown that Rab27a is required for the membrane docking of the secretory lysosome at the IS (Stinchcombe et al. 2001a). Whether similar mechanisms regulate the directed secretion of cytokines toward the helper T cell–APC interface is currently unknown.

The synapse assembly may also contribute to the arrival of intracellular pools of molecules regulating T cell activation. Antigenic activation of T cells is dramatically enhanced by the interaction of the cSMAC-localized costimulatory receptor CD28 with its ligands on the APC surface (Acuto and Michel 2003). Interestingly, CTLA-4, a transmembrane receptor retained in intracellular compartments in resting T cells, was shown to accumulate at the IS on TCR triggering (Egen and Allison 2002). CTLA-4 negatively regulate T cell activation by competing with CD28 for binding to its ligands (van der Merwe et al. 1997). The polarized delivery of CTLA-4 in areas where CD28 interacts with its ligands will inhibit CD28 costimulatory effects, therefore reducing T cell activation. TCR ligation induces the formation of dynamically regulated signaling complexes (Bunnell et al. 2002). Thus IS assembly could also be accompanied by the directional delivery of intracellular pools of signaling molecules including the adaptor protein LAT (Montoya et al. 2002) and the TCR complex itself (Das et al. 2004). The later event is regulated by SNAREs (Das et al. 2004), proteins specialized in vesicle docking and fusion during exocytosis (Bonifacino and Glick 2004), and its pharmacological inhibition limits TCR accumulation at the IS (Das et al. 2004). Thus the polarized delivery of proteins at the T cell–APC interface may allow the fine-tuning of T cell activation by regulating the composition of signaling complexes at the IS.

7.3 Protein Recycling and Degradation

Another possible role for the IS is to support protein recycling and/or degradation. This idea stems from the initial observation that, during T cell–APC interaction, several organelles are polarized toward the contact site (Kupfer and Dennert 1984; Kupfer et al. 1994; Peters et al. 1991). Importantly, whereas in resting cells TCRs are continuously internalized and recycled back to the cell surface (Alcover and Alarcon 2000; Liu et al. 2000), TCR ligation triggers an important downmodulation of the TCRs, involving reduced recycling and degradation by lysosomes and proteasomes (Liu et al. 2000). Together with the observation that TCR accumulation at the IS precedes its downmodulation (Lee et al. 2002), these data indicate that synapse formation regulates TCR recycling and degradation. Accordingly, T cell endosomes were found to be targeted to the IS on antigen-dependent T–APC interaction (Das et al. 2004; Le Bras et al. 2004), and disruption of tubulin cytoskeleton by colchicine blocks the polarization of recycling endosomes (Das et al. 2004) (see above). Further supporting the role of the cytoskeleton during receptor recycling at the IS, we have recently shown that the cytoplasmic adaptor HIP-55, a member of the drebrin/Abp1 family of actin-binding proteins (Lappalainen et al. 1998), localizes to the F-actin enriched T cell-APC contact site in an antigen-dependent manner (Le Bras et al. 2004). HIP-55 (also known as SH3P7) regulates distal signaling events in part through a specific downregulation of TCR expression. It is worth noting that HIP-55 was found in T cell early endocytic compartments translocating to the IS on TCR engagement (Le Bras et al. 2004), suggesting that HIP-55 may connect actin cytoskeleton and TCRs to endocytic processes. Interestingly, interfering with the expression of other components of T cell signaling has recently provided evidence of a connection between TCR trafficking/degradation and responsiveness to antigenic stimulation. For example, overexpression of Src-like adaptor protein-2 (SLAP-2) in Jurkat cells reduced TCR expression and NFAT activation (Loreto et al. 2002). Reduced trafficking of activated TCR in lymphocytes from c-Cbl x Cbl-b double-knockout mice has also been implicated in T cell hyperresponsiveness to TCR stimulation (Naramura et al. 2002). Members of the Cbl family of ubiquitin ligases ubiquitinate ZAP-70 and CD3 subunits, targeting the TCR complex for internalization and degradation (Rao et al. 2002). Genetic inactivation of the adaptor CD2AP, a member of the CIN85/CMS family of Cbl-interacting adaptors (Dikic and Giordano 2003), impairs TCR degradation after antigenic stimulation, leading to IL-2 hyperproduction and increased proliferation (Lee et al. 2003). It is worth noting here that the formation of a mature synapse was impaired in *CD2AP*$^{-/-}$ T cells, further indicating that IS is not required for T cell activation, but rather required for T cell desensitization. The fact that CIN85/CMS and drebrin/Abp1 families interact with actin cytoskeleton (Dikic and Giordano 2003; Lappalainen et al. 1998) indicates that a major function of Rho GTPase-dependent actin remodeling may be the regulation of protein recycling and degradation at the T cell–APC interface.

7.4 Transfer of Material Between the Two Cells

Another function of the IS may be the transfer of material between the two cells. For example, during the IS formation between CTLs and target cells, CTLs progressively integrate membrane markers from their targets (Huang et al. 1999; Stinchcombe et al. 2001). As a result, the CTL is transformed into a target itself, leading to a possible killing by neighboring fraticide T cells (Huang et al. 1999), thus turning down the cytotoxic response. Exchange of molecules between interacting cells has been shown in other models such as NK (Carlin et al. 2001), B (Batista et al. 2001), and helper T cells (Hwang et al. 2000). The mechanisms underlying this phenomenon are unclear. They could involve receptor endocytosis (Huang et al. 1999), exosomes (Blanchard et al. 2002b), or the formation of membrane bridges (Stinchcombe et al. 2001b). Although it is currently unknown whether Rho GTPases participate to intercellular transfer of material, their implication appears likely because actin cytoskeleton regulates a broad range of membrane-related processes including membrane protrusions, endocytosis, and lipid microdomain organization.

8 What About the APC?

One aspect of immunological synapse formation that is often overlooked is the role of the APC. The use of particular experimental systems such as planar lipid bilayers as surrogate APC is in large part responsible for this (Grakoui et al. 1999). Moreover, Wülfing et al. reported that cytochalasin D treatment of the B lymphoma cell line used as APC did not affect the formation of the synaptic pattern whereas treatment of the T-cell blocked recognition (Wulfing and Davis 1998). These observations initially favored a "passive model" describing the role played by the APC during T cell activation. However, recent studies indicate that APCs may actively contribute to IS formation. Al-Alwan et al. showed that DCs, which have an unique ability to activate naive T cells, polarize filamentous actin and fascin, an actin-bundling protein, during clustering with T cells. This process was critical to both clustering and activation of resting T cells (Al-Alwan et al. 2001). Using a TCR transgenic system, the same group further demonstrated that reorganization of the DC actin cytoskeleton was highly dependent of the presence of specific MHC-peptide complexes (Al-Alwan et al. 2003). In this study, DC cytoskeletal rearrangement was induced by directional ligation of MHC class II molecules, suggesting that motor or cytoskeletal proteins may drive this process. In addition, a recent study has shown that clustering of MHC class II molecules on B cells resulted in the cocapping of GEM domains and filamentous actin at the site of T cell-B cell conjugation, and these events were blocked by treating B cells with latrunculin B, a drug disrupting actin cytoskeleton (Gordy et al. 2004). It should be kept in mind that APCs express the counterreceptors of the T cell costimulatory molecules segregating during IS formation. Thus one can ask whether the above processes reflect a decreased adhesiveness often observed in the absence of cytoskeletal dynamics or a direct inhibition of

APC cytoskeletal proteins directing receptor movement. Nevertheless, these observations challenge the "passive model" during which the APC cytoskeleton does not play any role in the IS formation. Finally, actin cytoskeleton and Rho GTPases may regulate the molecular pathways involved in Ag uptake and processing by the APC, therefore modulating its capacity to stimulate T cells. Indeed, the activity of Cdc42 has been shown to play a critical role during Ag processing and DC maturation (Garrett et al. 2000; Mellman and Steinman 2001), and DCs from patients affected by chronic myeloid leukemia (CML) display altered actin reorganization associated with reduced antigen processing (Dong et al. 2003). Future studies will help in identifying the exact cytoskeletal mechanisms involved in Ag presentation by both professional and nonprofessional APCs.

9 Concluding Remarks

The redistribution of receptors and signaling molecules into organized patterns found at cell–cell junctions appears to be a common feature of lymphocyte activation. Although some of the mechanisms underlying this process have been unraveled, key questions are still pending. In particular, how are these specific patterns of molecules built up, and what are the cellular organelles and molecular motors directing these movements? Most importantly, what are immunological synapses established for? A particular effort should be devoted to identifying the specialized proteins—such as Rab proteins—implicated in protein and membrane transport. Rho GTPases are generally recognized as crucial regulators of the actin cytoskeleton. However, their ability to modulate cell polarity and motility, microtubule dynamics, vesicular trafficking, and gene transcription most likely implicates them in the major, if not all, processes governing lymphocyte biology. Future studies using imaging technologies in living cells and genetic models should certainly help in deciphering the functions of Rho GTPases during the establishment of immunological synapses and immune responses.

Acknowledgements This work is supported by grants from INSERM, Association pour la Recherche sur le Cancer (ARC), and Ministère délégué à la recherche et aux nouvelles technologies.

References

Acuto, O., and D. Cantrell. 2000. T cell activation and the cytoskeleton. *Annu Rev Immunol* 18:165–84.

Acuto, O., and F. Michel. 2003. CD28-mediated co-stimulation: a quantitative support for TCR signalling. *Nat Rev Immunol* 3:939–51.

Aghazadeh, B., W.E. Lowry, X.Y. Huang, and M.K. Rosen. 2000. Structural basis for relief of autoinhibition of the Dbl homology domain of proto-oncogene Vav by tyrosine phosphorylation. *Cell* 102:625–33.

Al-Alwan, M.M., R.S. Liwski, S.M. Haeryfar, W.H. Baldridge, D.W. Hoskin, G. Rowden, and K.A. West. 2003. Cutting edge: dendritic cell actin cytoskeletal polarization during immunological synapse formation is highly antigen-dependent. *J Immunol* 171:4479–83.

Al-Alwan, M.M., G. Rowden, T.D. Lee, and K.A. West. 2001. The dendritic cell cytoskeleton is critical for the formation of the immunological synapse. *J Immunol* 166:1452–6.

Alcover, A., and B. Alarcon. 2000. Internalization and intracellular fate of TCR-CD3 complexes. *Crit Rev Immunol* 20:325–46.

Allenspach, E.J., P. Cullinan, J. Tong, Q. Tang, A.G. Tesciuba, J.L. Cannon, S.M. Takahashi, R. Morgan, J.K. Burkhardt, and A.I. Sperling. 2001. ERM-dependent movement of CD43 defines a novel protein complex distal to the immunological synapse. *Immunity* 15:739–50.

Altman, A., and M. Deckert. 1999. The function of small GTPases in signaling by immune recognition and other leukocyte receptors. *Adv Immunol* 72:1–101.

Anton, I.M., M.A. de la Fuente, T.N. Sims, S. Freeman, N. Ramesh, J.H. Hartwig, M.L. Dustin, and R.S. Geha. 2002. WIP deficiency reveals a differential role for WIP and the actin cytoskeleton in T and B cell activation. *Immunity* 16:193–204.

Ardouin, L., M. Bracke, A. Mathiot, S.N. Pagakis, T. Norton, N. Hogg, and V.L. Tybulewicz. 2003. Vav1 transduces TCR signals required for LFA-1 function and cell polarization at the immunological synapse. *Eur J Immunol* 33:790–7.

Bachmann, M.F., S. Mariathasan, D. Bouchard, D.E. Speiser, and P.S. Ohashi. 1997. Four types of Ca^{2+} signals in naive $CD8^{+}$ cytotoxic T cells after stimulation with T cell agonists, partial agonists and antagonists. *Eur J Immunol* 27:3414–9.

Baier, G., D. Telford, L. Giampa, K.M. Coggeshall, G. Baier-Bitterlich, N. Isakov, and A. Altman. 1993. Molecular cloning and characterization of PKC θ, a novel member of the protein kinase C (PKC) gene family expressed predominantly in hematopoietic cells. *J Biol Chem* 268:4997–5004.

Barreiro, O., M. Vicente-Manzanares, A. Urzainqui, M. Yanez-Mo, and F. Sanchez-Madrid. 2004. Interactive protrusive structures during leukocyte adhesion and transendothelial migration. *Front Biosci* 9:1849-63.

Barreiro, O., M. Yanez-Mo, J.M. Serrador, M.C. Montoya, M. Vicente-Manzanares, R. Tejedor, H. Furthmayr, and F. Sanchez-Madrid. 2002. Dynamic interaction of VCAM-1 and ICAM-1 with moesin and ezrin in a novel endothelial docking structure for adherent leukocytes. *J Cell Biol* 157:1233–45.

Batista, F.D., D. Iber, and M.S. Neuberger. 2001. B cells acquire antigen from target cells after synapse formation. *Nature* 411:489–94.

Beeson, C., J. Rabinowitz, K. Tate, I. Gutgemann, Y.H. Chien, P.P. Jones, M.M. Davis, and H.M. McConnell. 1996. Early biochemical signals arise from low affinity TCR-ligand reactions at the cell–cell interface. *J Exp Med* 184:777–82.

Bi, K., and A. Altman. 2001. Membrane lipid microdomains and the role of PKCθ in T cell activation. *Semin Immunol* 13:139–46.

Blanchard, N., V. Di Bartolo, and C. Hivroz. 2002a. In the immune synapse, ZAP-70 controls T cell polarization and recruitment of signaling proteins but not formation of the synaptic pattern. *Immunity* 17:389–99.

Blanchard, N., D. Lankar, F. Faure, A. Regnault, C. Dumont, G. Raposo, and C. Hivroz. 2002b. TCR activation of human T cells induces the production of exosomes bearing the TCR/CD3/ζ complex. *J Immunol* 168:3235–41.

Bonifacino, J.S., and B.S. Glick. 2004. The mechanisms of vesicle budding and fusion. *Cell* 116:153–66.

Bousso, P., N.R. Bhakta, R.S. Lewis, and E. Robey. 2002. Dynamics of thymocyte-stromal cell interactions visualized by two-photon microscopy. *Science* 296:1876–80.

Brakebusch, C., and R. Fassler. 2003. The integrin-actin connection, an eternal love affair. *EMBO J* 22:2324–33.

Bromley, S.K., D.A. Peterson, M.D. Gunn, and M.L. Dustin. 2000. Cutting edge: hierarchy of chemokine receptor and TCR signals regulating T cell migration and proliferation. *J Immunol* 165:15–9.

Bubeck Wardenburg, J., R. Pappu, J.Y. Bu, B. Mayer, J. Chernoff, D. Straus, and A.C. Chan. 1998. Regulation of PAK activation and the T cell cytoskeleton by the linker protein SLP-76. *Immunity* 9:607–16.

Bunnell, S.C., D.I. Hong, J.R. Kardon, T. Yamazaki, C.J. McGlade, V.A. Barr, and L.E. Samelson. 2002. T cell receptor ligation induces the formation of dynamically regulated signaling assemblies. *J Cell Biol* 158:1263–75.

Cahalan, M.D., I. Parker, S.H. Wei, and M.J. Miller. 2002. Two-photon tissue imaging: seeing the immune system in a fresh light. *Nat Rev Immunol* 2:872–80.

Campbell, D.J., G.F. Debes, B. Johnston, E. Wilson, and E.C. Butcher. 2003. Targeting T cell responses by selective chemokine receptor expression. *Semin Immunol* 15:277–86.

Campbell, J.J., J. Hedrick, A. Zlotnik, M.A. Siani, D.A. Thompson, and E.C. Butcher. 1998. Chemokines and the arrest of lymphocytes rolling under flow conditions. *Science* 279:381–4.

Cantrell, D. 2002. Protein kinase B (Akt) regulation and function in T lymphocytes. *Semin Immunol* 14:19–26.

Cantrell, D.A. 2003. GTPases and T cell activation. *Immunol Rev* 192:122–30.

Carlin, L.M., K. Eleme, F.E. McCann, and D.M. Davis. 2001. Intercellular transfer and supramolecular organization of human leukocyte antigen C at inhibitory natural killer cell immune synapses. *J Exp Med* 194:1507–17.

Chan, A.C., D.M. Desai, and A. Weiss. 1994. The role of protein tyrosine kinases and protein tyrosine phosphatases in T cell antigen receptor signal transduction. *Annu Rev Immunol* 12:555–92.

Charvet, C., P. Auberger, S. Tartare-Deckert, A. Bernard, and M. Deckert. 2002. Vav1 couples T cell receptor to serum response factor-dependent transcription via a MEK-dependent pathway. *J Biol Chem* 21:21.

Clark, R., and G.M. Griffiths. 2003. Lytic granules, secretory lysosomes and disease. *Curr Opin Immunol* 15:516–21.

Clark, R.H., J.C. Stinchcombe, A. Day, E. Blott, S. Booth, G. Bossi, T. Hamblin, E.G. Davies, and G.M. Griffiths. 2003. Adaptor protein 3-dependent microtubule-mediated movement of lytic granules to the immunological synapse. *Nat Immunol* 4:1111–20.

Coleman, M.L., C.J. Marshall, and M.F. Olson. 2004. RAS and RHO GTPases in G_1-phase cell-cycle regulation. *Nat Rev Mol Cell Biol* 5:355–66.
Collins, T.L., M. Deckert, and A. Altman. 1997. Views on Vav. *Immunol Today* 18:221–5.
Corre, I., M. Gomez, S. Vielkind, and D.A. Cantrell. 2001. Analysis of thymocyte development reveals that the GTPase RhoA is a positive regulator of T cell receptor responses in vivo. *J Exp Med* 194:903–14.
Costello, P.S., M. Gallagher, and D.A. Cantrell. 2002. Sustained and dynamic inositol lipid metabolism inside and outside the immunological synapse. *Nat Immunol* 3:1082–9.
Crespo, P., K.E. Schuebel, A.A. Ostrom, J.S. Gutkind, and X.R. Bustelo. 1997. Phosphotyrosine-dependent activation of Rac-1 GDP/GTP exchange by the *vav* proto-oncogene product. *Nature* 385:169–172.
Das, V., B. Nal, A. Dujeancourt, M.I. Thoulouze, T. Galli, P. Roux, A. Dautry-Varsat, and A. Alcover. 2004. Activation-induced polarized recycling targets T cell antigen receptors to the immunological synapse; involvement of SNARE complexes. *Immunity* 20:577–88.
Davis, D.M. 2002. Assembly of the immunological synapse for T cells and NK cells. *Trends Immunol* 23:356-63.
Deckert, M., S. Tartare-Deckert, C. Couture, T. Mustelin, and A. Altman. 1996. Functional and physical interactions of Syk family kinases with the Vav proto-oncogene product. *Immunity* 5:591–604.
del Pozo, M.A., P. Sanchez-Mateos, M. Nieto, and F. Sanchez-Madrid. 1995. Chemokines regulate cellular polarization and adhesion receptor redistribution during lymphocyte interaction with endothelium and extracellular matrix. Involvement of cAMP signaling pathway. *J Cell Biol* 131:495–508.
del Pozo, M.A., M. Vicente-Manzanares, R. Tejedor, J.M. Serrador, and F. Sanchez-Madrid. 1999. Rho GTPases control migration and polarization of adhesion molecules and cytoskeletal ERM components in T lymphocytes. *Eur J Immunol* 29:3609–20.
Delon, J., N. Bercovici, G. Raposo, R. Liblau, and A. Trautmann. 1998. Antigen-dependent and -independent Ca^{2+} responses triggered in T cells by dendritic cells compared with B cells. *J Exp Med* 188:1473–84.
Delon, J., K. Kaibuchi, and R.N. Germain. 2001. Exclusion of CD43 from the immunological synapse is mediated by phosphorylation-regulated relocation of the cytoskeletal adaptor moesin. *Immunity* 15:691–701.
Dikic, I., and S. Giordano. 2003. Negative receptor signalling. *Curr Opin Cell Biol* 15:128–35.
Dong, R., K. Cwynarski, A. Entwistle, F. Marelli-Berg, F. Dazzi, E. Simpson, J.M. Goldman, J.V. Melo, R.I. Lechler, I. Bellantuono, A. Ridley, and G. Lombardi. 2003. Dendritic cells from CML patients have altered actin organization, reduced antigen processing, and impaired migration. *Blood* 101:3560–7.
Donnadieu, E., G. Bismuth, and A. Trautmann. 1994. Antigen recognition by helper T cells elicits a sequence of distinct changes of their shape and intracellular calcium. *Curr Biol* 4:584–95.
Doody, G.M., S.E. Bell, E. Vigorito, E. Clayton, S. McAdam, R. Tooze, C. Fernandez, I.J. Lee, and M. Turner. 2001. Signal transduction through Vav-2 participates in humoral immune responses and B cell maturation. *Nat Immunol* 2:542–7.
Doody, G.M., D.D. Billadeau, E. Clayton, A. Hutchings, R. Berland, S. McAdam, P.J. Leibson, and M. Turner. 2000. Vav-2 controls NFAT-dependent transcription in B- but not T-lymphocytes. *EMBO J.* 19:6173–84.

Dustin, M.L., S.K. Bromley, Z. Kan, D.A. Peterson, and E.R. Unanue. 1997. Antigen receptor engagement delivers a stop signal to migrating T lymphocytes. *Proc Natl Acad Sci USA* 94:3909–13.

Dustin, M.L., and A.C. Chan. 2000. Signaling takes shape in the immune system. *Cell* 103:283–94.

Egen, J.G., and J.P. Allison. 2002. Cytotoxic T lymphocyte antigen-4 accumulation in the immunological synapse is regulated by TCR signal strength. *Immunity* 16:23–35.

Etienne-Manneville, S., and A. Hall. 2002. Rho GTPases in cell biology. *Nature* 420:629–35.

Faure, S., L.I. Salazar-Fontana, M. Semichon, V.L. Tybulewicz, G. Bismuth, A. Trautmann, R.N. Germain, and J. Delon. 2004. ERM proteins regulate cytoskeleton relaxation promoting T cell-APC conjugation. *Nat Immunol* 5:272–9.

Fischer, K.D., Y.Y. Kong, H. Nishina, K. Tedford, L.E. Marengere, I. Kozieradzki, T. Sasaki, M. Starr, G. Chan, S. Gardener, M.P. Nghiem, D. Bouchard, M. Barbacid, A. Bernstein, and J.M. Penninger. 1998. Vav is a regulator of cytoskeletal reorganization mediated by the T- cell receptor. *Curr Biol* 8:554–62.

Friedl, P., and E.B. Brocker. 2002. TCR triggering on the move: diversity of T-cell interactions with antigen presenting cells. *Immunol Rev* 186:83–9.

Fujikawa, K., A.V. Miletic, F.W. Alt, R. Faccio, T. Brown, J. Hoog, J. Fredericks, S. Nishi, S. Mildiner, S.L. Moores, J. Brugge, F.S. Rosen, and W. Swat. 2003. Vav1/2/3-null mice define an essential role for Vav family proteins in lymphocyte development and activation but a differential requirement in MAPK signaling in T and B cells. *J Exp Med* 198:1595–608.

Gaide, O., B. Favier, D.F. Legler, D. Bonnet, B. Brissoni, S. Valitutti, C. Bron, J. Tschopp, and M. Thome. 2002. CARMA1 is a critical lipid raft-associated regulator of TCR-induced NF-κB activation. *Nat Immunol* 3:836–43.

Garrett, W.S., L.M. Chen, R. Kroschewski, M. Ebersold, S. Turley, S. Trombetta, J.E. Galan, and I. Mellman. 2000. Developmental control of endocytosis in dendritic cells by Cdc42. *Cell* 102:325–34.

Gomez, M., D. Kioussis, and D.A. Cantrell. 2001. The GTPase Rac-1 controls cell fate in the thymus by diverting thymocytes from positive to negative selection. *Immunity* 15:703–13.

Gomez, M., V. Tybulewicz, and D.A. Cantrell. 2000. Control of pre-T cell proliferation and differentiation by the GTPase Rac-I. *Nat Immunol* 1:348–52.

Gordy, C., S. Mishra, and W. Rodgers. 2004. Visualization of antigen presentation by actin-mediated targeting of glycolipid-enriched membrane domains to the immune synapse of B cell APCs. *J Immunol* 172:2030–8.

Grakoui, A., S.K. Bromley, C. Sumen, M.M. Davis, A.S. Shaw, P.M. Allen, and M.L. Dustin. 1999. The immunological synapse: a molecular machine controlling T cell activation. *Science* 285:221–7.

Griffiths, E.K., C. Krawczyk, Y.Y. Kong, M. Raab, S.J. Hyduk, D. Bouchard, V.S. Chan, I. Kozieradzki, A.J. Oliveira-Dos-Santos, A. Wakeham, P.S. Ohashi, M.I. Cybulsky, C.E. Rudd, and J.M. Penninger. 2001. Positive regulation of T cell activation and integrin adhesion by the adapter Fyb/Slap. *Science* 293:2260–3.

Gupta, S., J.C. Fanzo, C. Hu, D. Cox, S.Y. Jang, A.E. Lee, S. Greenberg, and A.B. Pernis. 2003. T cell receptor engagement leads to the recruitment of IBP, a novel guanine nucleotide exchange factor, to the immunological synapse. *J Biol Chem* 278:43541–9.

Haddad, E., J.L. Zugaza, F. Louache, N. Debili, C. Crouin, K. Schwarz, A. Fischer, W. Vainchenker, and J. Bertoglio. 2001. The interaction between Cdc42 and WASP is required for SDF-1-induced T lymphocyte chemotaxis. *Blood* 97:33–8.

Hailman, E., W.R. Burack, A.S. Shaw, M.L. Dustin, and P.M. Allen. 2002. Immature $CD4^+CD8^+$ thymocytes form a multifocal immunological synapse with sustained tyrosine phosphorylation. *Immunity* 16:839–48.

Harriague, J., and G. Bismuth. 2002. Imaging antigen-induced PI3K activation in T cells. *Nat Immunol* 3:1090-6.

Haston, W.S., J.M. Shields, and P.C. Wilkinson. 1982. Lymphocyte locomotion and attachment on two-dimensional surfaces and in three-dimensional matrices. *J Cell Biol* 92:747–52.

Hogg, N., M. Laschinger, K. Giles, and A. McDowall. 2003. T-cell integrins: more than just sticking points. *J Cell Sci* 116:4695–705.

Holsinger, L.J., I.A. Graef, W. Swat, T. Chi, D.M. Bautista, L. Davidson, R.S. Lewis, F.W. Alt, and G.R. Crabtree. 1998. Defects in actin-cap formation in Vav-deficient mice implicate an actin requirement for lymphocyte signal transduction. *Curr Biol* 8:563–72.

Huang, J.F., Y. Yang, H. Sepulveda, W. Shi, I. Hwang, P.A. Peterson, M.R. Jackson, J. Sprent, and Z. Cai. 1999. TCR-mediated internalization of peptide-MHC complexes acquired by T cells. *Science* 286:952–4.

Huby, R.D., G.W. Carlile, and S.C. Ley. 1995. Interactions between the protein-tyrosine kinase ZAP-70, the proto-oncoprotein Vav, and tubulin in Jurkat T cells. *J Biol Chem* 270:30241–4.

Huppa, J.B., and M.M. Davis. 2003. T-cell-antigen recognition and the immunological synapse. *Nat Rev Immunol* 3:973–83.

Huppa, J.B., M. Gleimer, C. Sumen, and M.M. Davis. 2003. Continuous T cell receptor signaling required for synapse maintenance and full effector potential. *Nat Immunol* 4:749–55.

Hwang, I., J.F. Huang, H. Kishimoto, A. Brunmark, P.A. Peterson, M.R. Jackson, C.D. Surh, Z. Cai, and J. Sprent. 2000. T cells can use either T cell receptor or CD28 receptors to absorb and internalize cell surface molecules derived from antigen-presenting cells. *J Exp Med* 191:1137–48.

Iezzi, G., K. Karjalainen, and A. Lanzavecchia. 1998. The duration of antigenic stimulation determines the fate of naive and effector T cells. *Immunity* 8:89–95.

Itoh, K., M. Sakakibara, S. Yamasaki, A. Takeuchi, H. Arase, M. Miyazaki, N. Nakajima, M. Okada, and T. Saito. 2002. Cutting edge: negative regulation of immune synapse formation by anchoring lipid raft to cytoskeleton through Cbp-EBP50-ERM assembly. *J Immunol* 168:541–4.

Ivetic, A., J. Deka, A. Ridley, and A. Ager. 2002. The cytoplasmic tail of L-selectin interacts with members of the ezrin-radixin-moesin (ERM) family of proteins: cell activation-dependent binding of moesin but not ezrin. *J Biol Chem* 277:2321–9.

Jackman, J.K., D.G. Motto, Q. Sun, M. Tanemoto, C.W. Turck, G.A. Peltz, G.A. Koretzky, and P.R. Findell. 1995. Molecular cloning of SLP-76, a 76-kDa tyrosine phosphoprotein associated with Grb2 in T cells. *J Biol Chem* 270:7029–7033.

Johnson, K.G., S.K. Bromley, M.L. Dustin, and M.L. Thomas. 2000. A supramolecular basis for CD45 tyrosine phosphatase regulation in sustained T cell activation. *Proc Natl Acad Sci USA* 97:10138–43.

Krause, M., A.S. Sechi, M. Konradt, D. Monner, F.B. Gertler, and J. Wehland. 2000. Fyn-binding protein (Fyb)/SLP-76-associated protein (SLAP), Ena/vasodilator-stimulated phosphoprotein (VASP) proteins and the Arp2/3 complex link T cell receptor (TCR) signaling to the actin cytoskeleton. *J Cell Biol* 149:181–94.

Krawczyk, C., A. Oliveira-dos-Santos, T. Sasaki, E. Griffiths, P.S. Ohashi, S. Snapper, F. Alt, and J.M. Penninger. 2002. Vav1 controls integrin clustering and MHC/peptide-specific cell adhesion to antigen-presenting cells. *Immunity* 16:331–43.

Krummel, M.F., M.D. Sjaastad, C. Wulfing, and M.M. Davis. 2000. Differential clustering of CD4 and CD3ζduring T cell recognition. *Science* 289:1349–52.

Kuhn, J.R., and M. Poenie. 2002. Dynamic polarization of the microtubule cytoskeleton during CTL-mediated killing. *Immunity* 16:111–21.

Kuhne, M.R., J. Lin, D. Yablonski, M.N. Mollenauer, L.I. Ehrlich, J. Huppa, M.M. Davis, and A. Weiss. 2003. Linker for activation of T cells, ζ-associated protein-70, and Src homology 2 domain-containing leukocyte protein-76 are required for TCR-induced microtubule-organizing center polarization. *J Immunol* 171:860–6.

Kupfer, A., and G. Dennert. 1984. Reorientation of the microtubule-organizing center and the Golgi apparatus in cloned cytotoxic lymphocytes triggered by binding to lysable target cells. *J Immunol* 133:2762–6.

Kupfer, A., and H. Kupfer. 2003. Imaging immune cell interactions and functions: SMACs and the Immunological Synapse. *Semin Immunol* 15:295–300.

Kupfer, A., T.R. Mosmann, and H. Kupfer. 1991. Polarized expression of cytokines in cell conjugates of helper T cells and splenic B cells. *Proc Natl Acad Sci USA* 88:775–9.

Kupfer, A., and S.J. Singer. 1989. Cell biology of cytotoxic and helper T cell functions: immunofluorescence microscopic studies of single cells and cell couples. *Annu Rev Immunol* 7:309–37.

Kupfer, A., S.L. Swain, and S.J. Singer. 1987. The specific direct interaction of helper T cells and antigen-presenting B cells. II. Reorientation of the microtubule organizing center and reorganization of the membrane-associated cytoskeleton inside the bound helper T cells. *J Exp Med* 165:1565–80.

Kupfer, H., C.R. Monks, and A. Kupfer. 1994. Small splenic B cells that bind to antigen-specific T helper (Th) cells and face the site of cytokine production in the Th cells selectively proliferate: immunofluorescence microscopic studies of Th-B antigen-presenting cell interactions. *J Exp Med* 179:1507–15.

Lang, P., L. Guizani, I. Vitte-Mony, R. Stancou, O. Dorseuil, G. Gacon, and J. Bertoglio. 1992. ADP ribosylation of the ras-related, GTP-binding protein RhoA inhibits lymphocyte-mediated cytotoxicity. *J Biol Chem* 267:11677–80.

Lappalainen, P., M.M. Kessels, M.J. Cope, and D.G. Drubin. 1998. The ADF homology (ADF-H) domain: a highly exploited actin-binding module. *Mol Biol Cell* 9:1951–9.

Le Bras, S., I. Foucault, A. Foussat, C. Brignone, O. Acuto, and M. Deckert. 2004. Recruitment of the actin binding protein HIP-55 to the immunological synapse regulates T cell receptor signaling and endocytosis. *J Biol Chem* 279:15550–60.

Lee, K.H., A.R. Dinner, C. Tu, G. Campi, S. Raychaudhuri, R. Varma, T.N. Sims, W.R. Burack, H. Wu, J. Wang, O. Kanagawa, M. Markiewicz, P.M. Allen, M.L. Dustin, A.K. Chakraborty, and A.S. Shaw. 2003. The immunological synapse balances T cell receptor signaling and degradation. *Science*

Lee, K.H., A.D. Holdorf, M.L. Dustin, A.C. Chan, P.M. Allen, and A.S. Shaw. 2002. T cell receptor signaling precedes immunological synapse formation. *Science* 295:1539–42.

Ley, S.C., W. Verbi, D.J. Pappin, B. Druker, A.A. Davies, and M.J. Crumpton. 1994. Tyrosine phosphorylation of αtubulin in human T lymphocytes. *Eur J Immunol* 24:99–106.

Liu, H., M. Rhodes, D.L. Wiest, and D.A. Vignali. 2000. On the dynamics of TCR:CD3 complex cell surface expression and downmodulation. *Immunity* 13:665–75.

Loreto, M.P., D.M. Berry, and C.J. McGlade. 2002. Functional cooperation between c-Cbl and Src-like adaptor protein 2 in the negative regulation of T-cell receptor signaling. *Mol Cell Biol* 22:4241–55.

Lowin-Kropf, B., V.S. Shapiro, and A. Weiss. 1998. Cytoskeletal polarization of T cells is regulated by an immunoreceptor tyrosine-based activation motif-dependent mechanism. *J Cell Biol* 140:861–71.

Luscinskas, F.W., S. Ma, A. Nusrat, C.A. Parkos, and S.K. Shaw. 2002. The role of endothelial cell lateral junctions during leukocyte trafficking. *Immunol Rev* 186:57–67.

Marie-Cardine, A., H. Kirchgessner, C. Eckerskorn, S.C. Meuer, and B. Schraven. 1995. Human T lymphocyte activation induces tyrosine phosphorylation of α-tubulin and its association with the SH2 domain of the p59fyn protein tyrosine kinase. *Eur J Immunol* 25:3290–7.

Matthews, S.A., E. Rozengurt, and D. Cantrell. 2000. Protein kinase D. A selective target for antigen receptors and a downstream target for protein kinase C in lymphocytes. *J Exp Med* 191:2075–82.

McEver, R.P. 2002. Selectins: lectins that initiate cell adhesion under flow. *Curr Opin Cell Biol* 14:581–6.

McFarland, W. 1969. Microspikes on the lymphocyte uropod. *Science* 163:818–20.

Mellman, I., and R.M. Steinman. 2001. Dendritic cells: specialized and regulated antigen processing machines. *Cell.* 106:255–8.

Menasche, G., E. Pastural, J. Feldmann, S. Certain, F. Ersoy, S. Dupuis, N. Wulffraat, D. Bianchi, A. Fischer, F. Le Deist, and G. de Saint Basile. 2000. Mutations in RAB27A cause Griscelli syndrome associated with haemophagocytic syndrome. *Nat Genet* 25:173–6.

Monks, C.R., B.A. Freiberg, H. Kupfer, N. Sciaky, and A. Kupfer. 1998. Three-dimensional segregation of supramolecular activation clusters in T cells. *Nature* 395:82–6.

Monks, C.R., H. Kupfer, I. Tamir, A. Barlow, and A. Kupfer. 1997. Selective modulation of protein kinase Cθ during T-cell activation. *Nature* 385:83–6.

Montoya, M.C., D. Sancho, G. Bonello, Y. Collette, C. Langlet, H.T. He, P. Aparicio, A. Alcover, D. Olive, and F. Sanchez-Madrid. 2002. Role of ICAM-3 in the initial interaction of T lymphocytes and APCs. *Nat Immunol* 3:159–68.

Moser, B., and P. Loetscher. 2001. Lymphocyte traffic control by chemokines. *Nat Immunol* 2:123–8.

Na, S., B. Li, I.S. Grewal, H. Enslen, R.J. Davis, J.H. Hanke, and R.A. Flavell. 1999. Expression of activated CDC42 induces T cell apoptosis in thymus and peripheral lymph organs via different pathways. *Oncogene* 18:7966–74.

Naramura, M., I.K. Jang, H. Kole, F. Huang, D. Haines, and H. Gu. 2002. c-Cbl and Cbl-b regulate T cell responsiveness by promoting ligand-induced TCR down-modulation. *Nat Immunol* 3:1192–9.

Nieto, M., J.M. Frade, D. Sancho, M. Mellado, A.C. Martinez, and F. Sanchez-Madrid. 1997. Polarization of chemokine receptors to the leading edge during lymphocyte chemotaxis. *J Exp Med* 186:153–8.

Ottoson, N.C., J.T. Pribila, A.S. Chan, and Y. Shimizu. 2001. Cutting edge: T cell migration regulated by CXCR4 chemokine receptor signaling to ZAP-70 tyrosine kinase. *J Immunol* 167:1857–61.

Palazzo, A.F., C.H. Eng, D.D. Schlaepfer, E.E. Marcantonio, and G.G. Gundersen. 2004. Localized stabilization of microtubules by integrin- and FAK-facilitated Rho signaling. *Science* 303:836–9.

Peters, P.J., J. Borst, V. Oorschot, M. Fukuda, O. Krahenbuhl, J. Tschopp, J.W. Slot, and H.J. Geuze. 1991. Cytotoxic T lymphocyte granules are secretory lysosomes, containing both perforin and granzymes. *J Exp Med* 173:1099–109.

Peterson, E.J. 2003. The TCR ADAPts to integrin-mediated cell adhesion. *Immunol Rev* 192:113–21.

Peterson, E.J., M.L. Woods, S.A. Dmowski, G. Derimanov, M.S. Jordan, J.N. Wu, P.S. Myung, Q.H. Liu, J.T. Pribila, B.D. Freedman, Y. Shimizu, and G.A. Koretzky. 2001. Coupling of the TCR to integrin activation by Slap-130/Fyb. *Science* 293:2263–5.

Pomerantz, J.L., E.M. Denny, and D. Baltimore. 2002. CARD11 mediates factor-specific activation of NFκB by the T cell receptor complex. *EMBO J.* 21:5184–94.

Potter, T.A., K. Grebe, B. Freiberg, and A. Kupfer. 2001. Formation of supramolecular activation clusters on fresh ex vivo $CD8^+$ T cells after engagement of the T cell antigen receptor and CD8 by antigen presenting cells. *Proc Natl Acad Sci USA.* 98:12624–9.

Rao, N., I. Dodge, and H. Band. 2002. The Cbl family of ubiquitin ligases: critical negative regulators of tyrosine kinase signaling in the immune system. *J Leukoc Biol* 71:753–63.

Revy, P., M. Sospedra, B. Barbour, and A. Trautmann. 2001. Functional antigen-independent synapses formed between T cells and dendritic cells. *Nat Immunol* 2:925–31.

Reynolds, L.F., L.A. Smyth, T. Norton, N. Freshney, J. Downward, D. Kioussis, and V.L. Tybulewicz. 2002. Vav1 transduces T cell receptor signals to the activation of phospholipase C-γ1 via phosphoinositide 3-kinase-dependent and -independent pathways. *J Exp Med* 195:1103–14.

Sahai, E., and C.J. Marshall. 2002. Rho-GTPases and cancer. *Nature Rev* 2:133–139.

Samelson, L.E. 2002. Signal transduction mediated by the T cell antigen receptor: the role of adapter proteins. *Annu Rev Immunol* 20:371–94.

Sanchez-Madrid, F., and M.A. del Pozo. 1999. Leukocyte polarization in cell migration and immune interactions. *EMBO J.* 18:501–11.

Sanchez-Martin, L., N. Sanchez-Sanchez, M.D. Gutierrez-Lopez, A.I. Rojo, M. Vicente-Manzanares, M.J. Perez-Alvarez, P. Sanchez-Mateos, X.R. Bustelo, A. Cuadrado, F. Sanchez-Madrid, J.L. Rodriguez-Fernandez, and C. Cabanas. 2004. Signaling through the leukocyte integrin LFA-1 in T cells induces a transient activation of Rac-1 that is regulated by Vav and PI3K/Akt-1. *J Biol Chem* 279:16194–205.

Serrador, J.M., J.R. Cabrero, D. Sancho, M. Mittelbrunn, A. Urzainqui, and F. Sanchez-Madrid. 2004. HDAC6 deacetylase activity links the tubulin cytoskeleton with immune synapse organization. *Immunity* 20:417–28.

Shaw, A.S., and M.L. Dustin. 1997. Making the T cell receptor go the distance: a topological view of T cell activation. *Immunity* 6:361–9.

Smith, A., M. Bracke, B. Leitinger, J.C. Porter, and N. Hogg. 2003. LFA-1-induced T cell migration on ICAM-1 involves regulation of MLCK-mediated attachment and ROCK-dependent detachment. *J Cell Sci* 116:3123–33.

Snapper, S.B., F.S. Rosen, E. Mizoguchi, P. Cohen, W. Khan, C.H. Liu, T.L. Hagemann, S.P. Kwan, R. Ferrini, L. Davidson, A.K. Bhan, and F.W. Alt. 1998. Wiskott-Aldrich syndrome protein-deficient mice reveal a role for WASP in T but not B cell activation. *Immunity* 9:81–91.

Stinchcombe, J.C., D.C. Barral, E.H. Mules, S. Booth, A.N. Hume, L.M. Machesky, M.C. Seabra, and G.M. Griffiths. 2001a. Rab27a is required for regulated secretion in cytotoxic T lymphocytes. *J Cell Biol* 152:825–34.

Stinchcombe, J.C., G. Bossi, S. Booth, and G.M. Griffiths. 2001b. The immunological synapse of CTL contains a secretory domain and membrane bridges. *Immunity* 15:751–61.

Stowers, L., D. Yelon, L.J. Berg, and J. Chant. 1995. Regulation of the polarization of T cells toward antigen-presenting cells by Ras-related GTPase CDC42. *Proc Natl Acad Sci USA* 92:5027–31.

Tanaka, Y., K. Bi, R. Kitamura, S. Hong, Y. Altman, A. Matsumoto, H. Tabata, S. Lebedeva, P.J. Bushway, and A. Altman. 2003. SWAP-70-like adapter of T cells, an adapter protein that regulates early TCR initiated signaling in Th2 lineage cells. *Immunity* 18:403–14.

Tartare-Deckert, S., M.N. Monthouel, C. Charvet, I. Foucault, E. Van Obberghen, A. Bernard, A. Altman, and M. Deckert. 2001. Vav2 activates c-fos serum response element and CD69 expression, but negatively regulates NF-AT and IL-2 gene activation in T lymphocyte. *J Biol Chem* 21:21.

Tedford, K., L. Nitschke, I. Girkontaite, A. Charlesworth, G. Chan, V. Sakk, M. Barbacid, and K.D. Fischer. 2001. Compensation between Vav-1 and Vav-2 in B cell development and antigen receptor signaling. *Nat Immunol* 2:548–55.

Ticchioni, M., C. Charvet, N. Noraz, L. Lamy, M. Steinberg, A. Bernard, and M. Deckert. 2002. Signaling through ZAP-70 is required for CXCL12-mediated T-cell transendothelial migration. *Blood* 99:3111–8.

Tomlinson, M.G., J. Lin, and A. Weiss. 2000. Lymphocytes with a complex: adapter proteins in antigen receptor signaling. *Immunol Today* 21:584–91.

Trautmann, A., and S. Valitutti. 2003. The diversity of immunological synapses. *Curr Opin Immunol* 15:249–54.

Turner, M., and D.D. Billadeau. 2002. VAV proteins as signal integrators for multi-subunit immune-recognition receptors. *Nat Rev Immunol* 2:476–86.

Urzainqui, A., J.M. Serrador, F. Viedma, M. Yanez-Mo, A. Rodriguez, A.L. Corbi, J.L. Alonso-Lebrero, A. Luque, M. Deckert, J. Vazquez, and F. Sanchez-Madrid. 2002. ITAM-based interaction of ERM proteins with Syk mediates signaling by the leukocyte adhesion receptor PSGL-1. *Immunity* 17:401–12.

Valitutti, S., M. Dessing, K. Aktories, H. Gallati, and A. Lanzavecchia. 1995. Sustained signaling leading to T cell activation results from prolonged T cell receptor occupancy. Role of T cell actin cytoskeleton. *J Exp Med* 181:577–84.

van der Merwe, P.A., D.L. Bodian, S. Daenke, P. Linsley, and S.J. Davis. 1997. CD80 (B7-1) binds both CD28 and CTLA-4 with a low affinity and very fast kinetics. *J Exp Med* 185:393–403.

Vicente-Manzanares, M., J.R. Cabrero, M. Rey, M. Perez-Martinez, A. Ursa, K. Itoh, and F. Sanchez-Madrid. 2002. A role for the Rho-p160 Rho coiled-coil kinase axis in the chemokine stromal cell-derived factor-1α-induced lymphocyte actomyosin and microtubular organization and chemotaxis. *J Immunol* 168:400–10.

Vicente-Manzanares, M., and F. Sanchez-Madrid. 2004. Role of the cytoskeleton during leukocyte responses. *Nat Rev Immunol* 4:110–22.

Villalba, M., K. Bi, J. Hu, Y. Altman, P. Bushway, E. Reits, J. Neefjes, G. Baier, R.T. Abraham, and A. Altman. 2002. Translocation of PKCθ in T cells is mediated by a non-conventional, PI3-K- and Vav dependent pathway, but does not absolutely require phospholipase C. *J Cell Biol* 157:253–63.

Villalba, M., K. Bi, F. Rodriguez, Y. Tanaka, S. Schoenberger, and A. Altman. 2001. Vav1/Rac-dependent actin cytoskeleton reorganization is required for lipid raft clustering in T cells. *J Cell Biol* 155:331–8.

Villalba, M., N. Coudronniere, M. Deckert, E. Teixeiro, P. Mas, and A. Altman. 2000. A novel functional interaction between Vav and PKCθ is required for TCR-induced T cell activation. *Immunity* 12:151–60.

Wang, D., R. Matsumoto, Y. You, T. Che, X.Y. Lin, S.L. Gaffen, and X. Lin. 2004. CD3/CD28 costimulation-induced NF-κB activation is mediated by recruitment of protein kinase C-θ, Bcl10, and IκB kinase βto the immunological synapse through CARMA1. *Mol Cell Biol* 24:164–71.
Wulfing, C., and M.M. Davis. 1998. A receptor/cytoskeletal movement triggered by costimulation during T cell activation. *Science* 282:2266–9.
Wulfing, C., J.D. Rabinowitz, C. Beeson, M.D. Sjaastad, H.M. McConnell, and M.M. Davis. 1997. Kinetics and extent of T cell activation as measured with the calcium signal. *J Exp Med* 185:1815–25.
Yonemura, S., and S. Tsukita. 1999. Direct involvement of ezrin/radixin/moesin (ERM)-binding membrane proteins in the organization of microvilli in collaboration with activated ERM proteins. *J Cell Biol* 145:1497–509.
Yu, H., D. Leitenberg, B. Li, and R.A. Flavell. 2001. Deficiency of small GTPase Rac2 affects T cell activation. *J Exp Med* 194:915–26.
Zakaria, S., T.S. Gomez, D.N. Savoy, S. McAdam, M. Turner, R.T. Abraham, and D.D. Billadeau. 2004. Differential regulation of TCR-mediated gene transcription by Vav family members. *J Exp Med* 199:429–34.
Zeng, R., J.L. Cannon, R.T. Abraham, M. Way, D.D. Billadeau, J. Bubeck-Wardenberg, and J.K. Burkhardt. 2003. SLP-76 coordinates Nck-dependent Wiskott-Aldrich syndrome protein recruitment with Vav-1/Cdc42-dependent Wiskott-Aldrich syndrome protein activation at the T cell-APC contact site. *J Immunol* 171:1360–8.
Zhang, W., J. Sloan-Lancaster, J. Kitchen, R.P. Trible, and L.E. Samelson. 1998. LAT: the ZAP-70 tyrosine kinase substrate that links T cell receptor to cellular activation. *Cell* 92:83–92.

CTMI (2005) 291:91–111

Rho GTPases and the Control of the Oxidative Burst in Polymorphonuclear Leukocytes

B. A. Diebold · G. M. Bokoch (✉)

Departments of Immunology and Cell Biology, The Scripps Research Institute, IMM-14, 10550 N. Torrey Pines Road, La Jolla CA, 92037, USA
bokoch@scripps.edu

Abstract Stimulation of quiescent leukocytes activates the NADPH oxidase, a membrane-associated enzyme system that generates superoxide and other reactive oxygen species (ROS) that are used to kill bacteria within the phagosome. This chapter describes this multicomponent NADPH oxidase system, one of the first cellular systems shown to be directly regulated by Rac GTPases. We present current models of NADPH oxidase regulation by Rac2 and describe how Rac2 activation controls the timing of ROS production in adherent neutrophils. The antagonistic role that Cdc42 plays as a competitor of Rac2 for binding to the cytochrome component of the NADPH oxidase is discussed as a possible mechanism for tonic regulation of ROS production during the formation of the phagosome. Finally, we briefly depict mechanisms by which invasive bacteria can alter (inhibit) NADPH oxidase function, focusing on the effects of invasive bacteria on components and assembly of the NADPH oxidase.

Abbreviations

Arp	Actin-related protein
CFP	Cyanin fluorescent protein
CGD	Chronic granulomatous disease
CRIB	Cdc42 and Rac interactive binding domain
cyt b	Cytochrome b_{558}
fMLF	Formyl-methionyl leucyl phenylalanine
FRET	Fluorescence resonance energy transfer
GAP	GTPase activating protein
GDI	GDP dissociation inhibitor
GEF	Guanine nucleotide exchange factor
GST	Glutathione *S*-transferase
H_2O_2	Hydrogen peroxide
HOCl	Hypochlorous acid
IFN-γ	Interferon gamma
mant	Methylanthraniloyl
O_2^-	Superoxide anion
•OH	Hydroxyl radical
Pak	p21-Activated kinase
PBD	p21-Binding domain of Pak
phox	Phagocyte oxidase
PMA	Phorbol myristate acetate
PMN	Polymorphonuclear granulocytes
PtdIns(3)P	Phosphatidylinositol-3-phosphate
ROS	Reactive oxygen species
SH3	src Homology 3
SPI	*Salmonella* pathogenicity island
TPR	Tetratricopeptide
TTSS	Type III secretion system
WASp	Wiskott-Aldrich syndrome protein
YFP	Yellow fluorescent protein

1 Introduction

Invasive bacteria are not the passive victims of phagocytosis and the phagocyte killing machinery depicted in the past. They are, on the contrary, quite robust in defending themselves against the host's immune system, and they have developed elegant defense strategies to promote their survival (reviewed in Cornelis 2002; Cosart and Sansonetti 2004; Sibley 2004). As discussed in detail in other chapters in this volume, some bacteria have developed ingenious mechanisms to avert phagocytosis, whereas other bacteria promote their own uptake into the phagosome, where the bacteria can elude extracellular immunological surveillance. Whether the bacteria are localized inside or outside of the host cell, they produce an arsenal of virulence factors that affect cell function and enable bacterial survival. One mechanism by which bacteria subvert host cell function is by regulating Rho GTPases. This is an excellent defense strategy because Rho GTPases are critical to a broad range of signaling pathways controlling immune cell function (reviewed in Bishop and Hall 2000; Burridge and Wennerberg 2004; Dharmawardhane and Bokoch 1997; Matozaki et al. 2000). The weaponry used against the Rho GTPases (reviewed in Barbieri et al. 2002; Bliska 2000; Cornelis 2002; Stebins and Galan 2001) includes proteins that function as GTPase-activating proteins (GAPs), which inactivate Rho GTPases by stimulating GTP hydrolysis (i.e., *Yersinia*, Yop E), or as guanine nucleotide exchange factors (GEFs), which catalyze GTP-for-GDP exchange to directly activate Rho GTPases (i.e., *Salmonella*, SopE). Other bacterial proteins are tyrosine phosphatases that inactivate endogenous GEFs, thus preventing Rho GTPase activation (i.e., *Salmonella*, SptP; *Yersinia*, YopH). In addition, some virulence factors are proteases that directly degrade Rho GTPases (i.e., *Yersinia*, YopT), whereas others are toxins that can modify and inhibit Rho GTPases by ADP-ribosylation (*Clostridium botulinum* C3 exotransferase) or glucosylation (*Clostridium difficile* toxins A and B). Indeed, there exist other bacterial proteins that interact with Rho GTPases, but their exact regulatory actions have yet to be discovered (i.e., *Yersinia* YopO).

The purpose of this chapter is to provide a description of one of the best-understood immunological responses whose function is pivotally dependent on Rho GTPase activity, namely, the oxidative burst of phagocytic polymorphonuclear leukocytes (PMN) (reviewed in Babior 1999; Babior et al. 2002; Bokoch and Knaus 2003). After PMN have engulfed microorganisms, they generate toxic reactive oxygen species (ROS) within the lumen of the phagosome. These include superoxide anion (O_2^-), and other oxidants derived from it, particularly hydrogen peroxide (H_2O_2), hydroxyl radical (OH•) and hypochlorous acid (HOCl). Superoxide anion is produced by the action of the NADPH oxidase, which catalyzes the one electron-reduction of O_2 to O_2^- using NADPH as substrate. Thus the respiratory or oxidative burst is due to the sudden increase in O_2 consumption when PMN produce O_2^- via the NADPH oxidase.

The NADPH oxidase is dormant in quiescent cells but becomes rapidly activated by a variety of stimuli associated with leukocyte chemotaxis and phagocytosis. These include a number of biologically active lipids, soluble chemoattractant peptides (e.g., complement component C5a and formyl peptides produced as byproducts of bacterial protein secretion), and opsonized particulate stimuli, all of which bind to specific cell surface receptors present on neutrophils, macrophages, and eosinophils. NADPH oxidase was one of the first cellular systems shown to be directly regulated by a member of the Ras superfamily of GTP-binding proteins: Rac1 or Rac2 (Abo et al. 1991; Knaus et al. 1991). The availability of cell-free assays for reconstitution studies and knowledge of the protein components of this system have led to much insight into the mechanisms through which Rac regulates NADPH oxidase activity. This chapter will describe our current model of this regulatory mechanism. We will discuss the signaling pathways leading to Rac2 activation in adherent neutrophils and describe a mechanism for functionally antagonistic cross talk between Rac and Cdc42 GTPases in regulation of ROS production. Finally, we will briefly consider current knowledge of how bacterial pathogens can affect this system.

2 Components of the NADPH Oxidase

The NADPH oxidase is a multiprotein system that is composed of both membrane-bound and cytosolic components (Fig. 1). The electron transfer reactions catalyzed by the NADPH oxidase require the action of the membrane-bound cytochrome b_{558} (cyt b). Cyt b is a flavohemeprotein that is composed of two transmembrane subunits, gp91^{phox} and p22^{phox} (*ph*agocytic *ox*idase). In addition to the two cyt b heme groups, there are binding sites for NADPH and FAD on the gp91^{phox} C terminus. Although cyt b possesses the components to catalyze the electron transfer process, it cannot perform this function without interacting with the cytosolic oxidase components. Malfunction mutations in cyt b subunits or the cytosolic regulatory components, or their complete absence, results in the inherited disorder known as chronic granulomatous disease (CGD) (Heyworth et al. 2003). PMN of CGD patients are unable to mount an oxidative burst in response to infections, with an often lethal result.

p47^{phox} is one of the NADPH oxidase regulatory components, found in a preformed cytosolic complex with p67^{phox} and p40^{phox} in resting leukocytes. In the absence of stimulation, p47^{phox} is maintained in an inactive conformation by an intramolecular interaction between tandem SH3 domains, which form a "super"-SH3 domain, and a nonconventional C-terminal polyproline domain (Groemping et al. 2003). When PMN are activated, p47^{phox} is phosphorylated on multiple serine residues at the border of the intramolecular binding surface. The phosphorylation of p47^{phox} destabilizes the autoinhib-

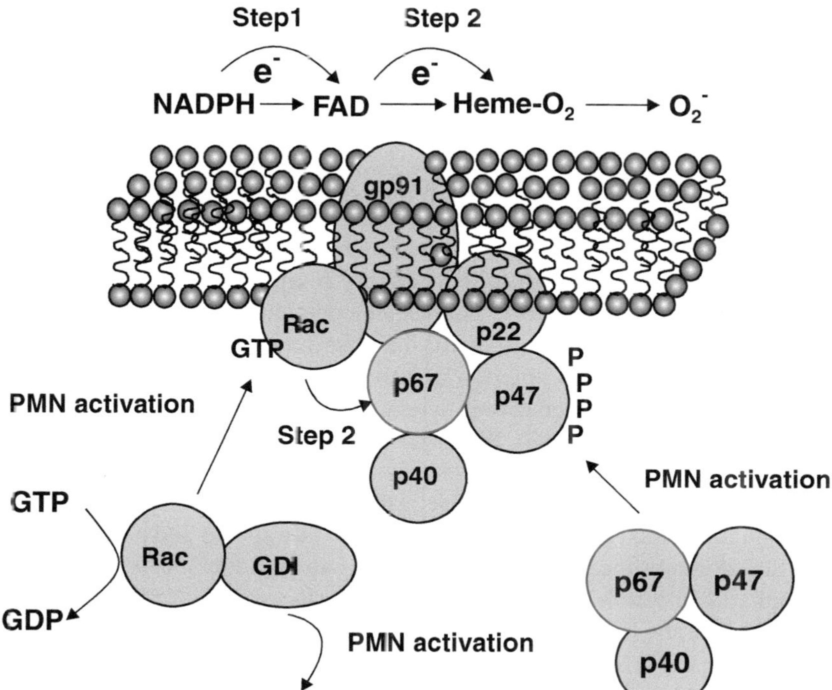

Fig. 1 Regulation of the phatocytic NADPH oxidase by Rac. Phosphorylation of p47^{phox} in the cytosol leads to translocation of the p47^{phox}-p67^{phox}-p40^{phox} complex to the phagosomal membrane, where p47^{phox} binds to p22^{phox} (the smaller subunit of flavocytochrome b_{558}) and acts as an adaptor for p67^{phox}, which binds to gp91^{phox} (the larger subunit of flavocytochrome b_{558}). gp91^{phox} contains the binding sites for NADPH and FAD. In addition to these events, Rac is released from GDI and is converted to its GTP-bound form. Rac-GTP translocates to the membrane simultaneously, but independently from the translocation of the p47^{phox}-p67^{phox}-p40^{phox} complex. At the membrane Rac interacts with phospholipids via its prenylated C-terminus and with cyt b via its insert domain. In *Step 1* of electron transfer, Rac and p67^{phox} do not interact, but their independent interactions with cytochrome b_{558} allow electrons to flow from NADPH to FAD. In *Step 2* of electron transfer, the Switch 1 domain of Rac engages the TPR domain of p67^{phox}, and this allows electrons to continue to flow from FAD to oxygen. The electrons from NADPH are used in the single electron reduction of oxygen, resulting in the production of superoxide anion (O_2^-)

ited conformation, enabling p47^{phox} to bind via the super-SH3 domain with enhanced affinity to a proline-rich motif on p22^{phox} (Groemping et al. 2003; Yuzawa et al. 2004). A PX ("phox") domain is also exposed, which mediates interactions with membrane-localized phosphoinositides, thereby promoting membrane translocation of p47^{phox} and the associated p67^{phox} (reviewed in Wientjes and Segal 2003). Thus, through this unique mechanism, p47^{phox} serves as both a regulatory response element for oxidase assembly induced by extracellular activators and an adapter to facilitate binding of p67^{phox} with cyt b. The latter role has been confirmed in cell-free NADPH oxidase assays, where the presence of p47^{phox} is unnecessary when higher concentrations of p67^{phox} and Rac are used (Freeman and Lambeth 1996; Koshin et al. 1996).

p67^{phox} is an essential cytosolic regulatory component of the NADPH oxidase. The middle of the p67^{phox} molecule contains a region that has been identified as an "activation domain," because mutation within this region leads to a loss of oxidase function in vitro (Han et al. 1999). The activation domain of p67^{phox} has been shown to regulate electron transfer from NADPH to FAD (Nisimoto et al. 1999). The N-terminus of p67^{phox} contains four tetratricopeptide (TPR) motifs that have been shown by X-ray crystallography to interact with the Switch 1 region of activated Rac (Lapouge et al. 2000). In the unstimulated PMN, p67^{phox} is bound to p47^{phox} (see above) and p40^{phox} via its two SH3 domains located in its C-terminus (Lapouge 2002). p40^{phox} binds to phosphatidylinositol-3-phosphate and translocates with p47^{phox} and p67^{phox} to the membrane (Ellson et al. 2001; Kanai et al 2001), but the significance of this finding is still unclear because p40^{phox} is not essential in cell-free assays.

3 Regulation of NADPH Oxidase by Rac GTPase

Stimulation of phagocytic PMN and the resulting phosphorylation of p47^{phox} trigger the translocation of the p47^{phox}-p67^{phox}-p40^{phox} complex from the cytosol to the plasma membrane, where p47^{phox} and p67^{phox} interact with cyt b. Cellular activation also induces Rac to separate from inert cytosolic complexes with GDP dissociation inhibitor (GDI), perhaps due to the phosphorylation of GDI by p21-activated kinase 1(Pak 1) (DerMardirossian et al. 2004) and the action of lipid mediators (Chuang et al. 1993). Specific GEFs (as yet unidentified) then catalyze the exchange of GDP for GTP on Rac. Rac translocates to the membrane simultaneously with, but independently of, the translocation of the 47^{phox}-p67^{phox}-p40^{phox} complex (Bokoch et al. 1994; Dorseuil et al. 1995; Heyworth et al. 1994; Quinn et al. 1993). The absolute requirement for Rac for NADPH oxidase activity (reviewed in Dinauer 2003) has been confirmed by the generation of Rac2-null mice (Roberts et al. 1999), through the use of Rac antisense oligonucleotides (Dorseuil et al. 1992), and by analysis of mice in which Bcr (a GTPase activating protein for Rac) has been knocked out (Voncken et al. 1995). Interestingly, whereas Rac2 is the major oxidase regulatory isoform in human neutrophils, Rac1 appears to play this role in human monocytes (Zhao et al. 2003).

Rac regulates phagocyte ROS production by a two-step mechanism in which Rac first directly interacts with cyt b independently of p67*phox* but acts in coordination with p67*phox* to allow electrons to flow from NADPH to FAD (Step 1) (Diebold and Bokoch 2001; Bokoch and Diebold 2002). Then, Rac must interact directly with p67*phox* in a second step to allow electrons to flow from FAD to oxygen (Step 2) to form O_2^-. This model is based on results obtained with a Rac2 Switch 1 mutant (Rac2 D38A) that could not bind p67*phox* and a p67*phox* construct lacking the ability to bind Rac. Using separate cell-free assays to measure electron flow either from NADPH to FAD (Step 1) or from NADPH to oxygen (Step 2), we showed that the noninteracting Rac2 and p67*phox* constructs allowed electron flow from NADPH to FAD (Step 1), but not from FAD to oxygen, to take place. This suggested that although both Rac and p67 were required for the Step 1 reaction, they did not have to interact with each other in this step. Their interaction, however, was required for the Step 2 reaction.

Using a fluorescent, nonhydrolyzable analog of GTP, 1′(3′)-*O*-(*N*-methyl-anthraniloyl)-GppNHp (mant-GppNHp), bound to Rac2, we demonstrated that Rac2 interacts with cyt b via the GTPase insert region (aa 124–136). Binding was observed when mant-GTP-Rac2WT was added to cyt b, but not when Rac2 Δ124–135 was used in place of Rac2 WT. Consistent with cyt b binding, this mutant did not support electron transfer in Step 1 (or Step 2). These data did not support prior dogma that Rac interacts solely with p67*phox* and serves as a docking protein to orient p67*phox*, the only regulator of electron flow between NADPH and FAD, in relation to its target, cyt b. These results established that Rac also contributes to the regulation of oxidase electron transfer and provide a reasonable explanation for prior observations that Rac GTPase is indispensable for NADPH oxidase function.

4 The Pivotal Role of Rac2 in Adhesion-Mediated Suppression of ROS in Neutrophils

In addition to being a regulatory component of electron flow of the NADPH oxidase, Rac2 serves as a common point of convergence for integrin and chemoattractant receptor cross talk in neutrophils (Zhao et al. 2003). Neutrophils that were adherent to fibronectin- or fibrinogen-coated tissue culture plates were observed to have an oxidative burst that was delayed by 30–90 min after stimulation with formyl-methionyl leucyl phenylalanine (fMLF) peptide or C5a. The delay has been shown to be a result of $\beta 1$ (fibronectin)- or $\beta 2$ (fibrinogen)-integrin engagement. The prolonged lag period in the oxidative response in neutrophils adherent to surfaces coated with extracellular matrix proteins is presumed to correspond to the in situ situation in which the cells are migrating through the tissue to inflammatory sites. After the cells reach these sites, the neutrophils then respond and generate microbicidal ROS.

Using the glutathione *S*-transferase-p21-binding domain of Pak1 (GST-PBD) fusion protein to pull down active Rac2 (Benard et al. 1999) from lysates of neutrophils stimulated with fMLF or C5a, we observed that Rac2 activation was also delayed by 30–90 min in adherent neutrophils, but not in suspended cells (Zhao et al. 2003). Studies on NADPH oxidase assembly showed that adhesion signals also resulted in delayed Rac2 translocation to the plasma membrane, but the translocation of the other cytosolic oxidase components, $p47^{phox}$ and $p67^{phox}$, occurred normally. These data suggested that integrin signaling regulates NADPH oxidase function by specifically controlling the activation and translocation of Rac2. This hypothesis was verified by demonstrating that the suppressive effects of adhesion on the oxidative burst could be reversed by introduction of recombinant, constitutively active Rac2 G12V into the adherent neutrophils.

The defect in Rac2 activation was localized to a membrane component by conducting cell-free NADPH oxidase assays in which membranes prepared from adherent or suspended neutrophils were reconstituted with cytosol from either adherent or suspended neutrophils in mix-match studies. Membranes from adherent neutrophils could not produce ROS, implicating a membrane component in the integrin-dependent defect. It appeared likely that a Rac GEF rather than an overactive GAP was responsible for the inactive state of Rac2 in adherent neutrophils, and, indeed, it was observed that activation of the membrane-associated Rac GEF Vav1 was inhibited in adherent cells. Vav1 phosphorylation on the critical Tyr 174 site was observed only when Rac2 activation and ROS formation was evident. Examination of Syk activation, which lies upstream of Vav1 and phosphorylates Vav1 on Tyr 174 (Aghazadeh et al. 2000; Moores et al 2000), revealed that Syk activation was normal and immediate on adherent neutrophil stimulation. The general tyrosine phosphatase inhibitor sodium orthovanadate reversed the effects of adhesion on the kinetics of the respiratory burst, Rac activation, and Vav1 activation, correlating with an increase in tyrosine phosphorylated proteins in adherent neutrophils, making the activation of one or more tyrosine phosphatases a likely signal for adhesion-mediated suppression of NADPH oxidase activation (Fig. 2). One scenario by which invasive bacteria could alter oxidant production would be to fool the neutrophils into an "adherent mode" in which a bacterially derived

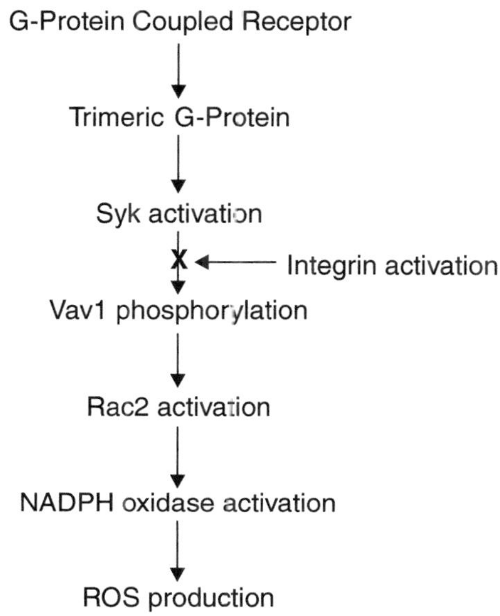

Fig. 2 Adhesion-mediated suppression of reactive oxygen species. Integrin activation by adherence of PMN to fibronectin or fibrinogen attenuates ROS production by the NADPH oxidase because of a delay in Vav1 activation and a consequent delay in Rac2 activation

protein phosphatase would prevent Vav1 and Rac2 activation, thus preventing ROS production. Indeed, many bacterial pathogens bind to cell adhesion molecules of host cells and alter signaling pathways to allow their survival (reviewed in Boyle and Finlay 2003).

5
Inhibitory Regulation of ROS Production by Cdc42

We recently reported that Cdc42 antagonizes Rac in the regulation of oxidant production by neutrophils (Diebold et al. 2004). Cdc42, unlike Rac, cannot support NADPH oxidase activity in the cell-free assay system because of two residues within the Switch I region required for p67^{phox} binding (Lapouge et al. 2000) that differ between Rac and Cdc42. Mutation of these residues in Cdc42 to the corresponding residues of Rac enabled Cdc42 (K27A, S30G) to now support O_2^- production to the same extent as Rac WT (Kwong et al. 1994). This indicated that the Cdc42 insert domain was indeed functional and suggested the possibility that Cdc42 might compete with Rac for binding cyt b under normal circumstances. We observed that GST-Cdc42, as well as GST-Rac1 and Rac2, could specifically bind cyt b in in vitro pull down assays (Diebold et al. 2004). This interaction was only partially GTP dependent but was dependent on the GTPase insert domain.

In NADPH oxidase cell-free assays, Cdc42 WT, but not Cdc42Δ124–135, inhibited Rac2(or Rac1)-induced superoxide production (Diebold et al. 2004). The inhibitory effect of Cdc42 WT was decreased when the concentration of Rac2 was increased in the assay, suggesting that Cdc42 competes with Rac2 for binding to cyt b via the insert domain. This hypothesis was confirmed in a direct competition binding assay in which decreasing amounts

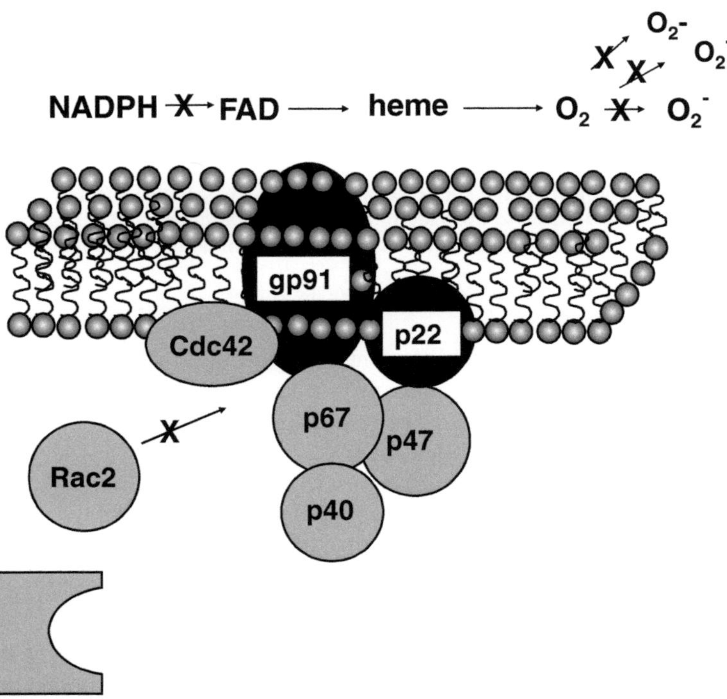

Fig. 3 Cdc42 inhibits Rac-induced ROS production by the phagocytic NADPH oxidase. Rac binds to cyt b via its insert domain. Cdc42 competes with Rac for cyt b binding and inhibits superoxide production in in vitro cell-free assays. These effects of Cdc42 are dependent on its insert domain. In vivo, overexpression of Cdc42, but not Cdc42Δinsert domain, inhibited Rac1Q61L-induced ROS production in a genetically engineered Cos cell line expressing NADPH oxidase components. In human neutrophils, sequestration of endogenous Cdc42, but not of Rac, by the expression of Wiskott-Aldrich syndrome protein (*WASp*) CRIB domain increased ROS production (see text for details)

of cyt b were bound by GST-Rac2 when increasing amounts of Cdc42 were present. In vivo experiments showed that Cdc42 could inhibit Rac1-induced ROS production in a genetically engineered Cos cell line that stably expresses a functional NADPH oxidase (Price et al. 2002). Using human neutrophils, we also showed that sequestration of endogenous Cdc42 by introduction of the Cdc42/Rac-interactive binding domain (CRIB) of Wiskott-Aldrich syndrome protein (WASp) into the cells increased O_2^- formation two- to threefold on stimulation with fMLF, providing further evidence that Cdc42 plays an antagonistic role in regulating Rac-induced ROS production (Fig. 3).

6 Timing Is Everything: Temporal Regulation of Rho GTPases During Phagocytosis and ROS Production

Cdc42 is activated nearly simultaneously with Rac2 in chemoattractant-stimulated human neutrophils (Benard et al 1999). Activation of Cdc42 is required for the cell polarization necessary for leukocyte chemotaxis, as well as for assembly of the motile actin machinery via the WASp-Arp2/3 complex (Li et al. 2003; Srinivasan et al. 2003). Cdc42 may thus serve as a tonic regulator to dampen the amount of ROS generated during leukocyte transmigration through tissues. Activation of Cdc42 might also inhibit full oxidant production until phagocytic cup formation and bacterial uptake are completed. This would coordinate ROS formation with the bacterial uptake process for the most efficient killing. A recent study on the timing of Rho GTPase activation during phagocytosis was conducted with FRET analysis (Hoppe and Swanson 2004). In RAW264.7 macrophages undergoing phagocytosis of opsonized erythrocytes, the YFP-Cdc42/CFP-PBD FRET signal reported activation of Cdc42 at the site of particle attachment, immediately after contact with the macrophage. Cdc42 was active only at the advancing edge of the pseudopod and remained active during the pseudopod extension phase of phagosome formation. Before phagosome closure, the activity of Cdc42 decreased. In contrast, YFP-Rac1 and CFP-PBD interaction was evident shortly after particle contact and was seen throughout the extending pseudopod. During the closure phase, active Rac1 was observed at the base of the phagosome. The activation pattern of Rac2 differed substantially. Rac 2 was only slightly active in the vicinity of the particle and displayed only minor activation during pseudopod extension. There was, however, a pronounced transient increase in Rac2 activity distributed over the base of the phagosome during closure. The presence of active Cdc42 and the absence of active Rac2 during the early phases of phagocytosis are consistent with the concept that Cdc42 may indeed play an antagonistic role with respect to NADPH oxidase activation by Rac2 during formation of the phagocytic cup. We note the possibility that by activating Cdc42, pathogenic bacteria could reduce the level of ROS in their phagosomal environment.

7 Reactive Oxygen Species and Proteases: The Warheads

Superoxide anion (O_2^-) and the ROS that are generated from it are toxic to engulfed pathogens (Roos et al. 2003). O_2^- is unstable and dismutates spontaneously in acidic environments to hydrogen peroxide (H_2O_2) and oxygen:

$$2\,O_2^- + 2\,H^+ \rightarrow H_2O_2 + O_2$$

The microbicidal potency of the H_2O_2 within the phagosome is greatly increased by the enzyme myeloperoxidase, which is supplied to the phagosome by the fusion of azurophilic granules. This enzyme catalyzes the oxidation of halide ions by H_2O_2 to form hypohalite ions, one of the principal classes of microbicidal agents produced within the phagosome. The reaction of H_2O_2 and chloride ion produces hypochlorite ion (^-OCl), which is protonated to hypochlorous acid (HOCl) at acidic pH:

$$H^+ + H_2O_2 + Cl^- \rightarrow HOCl + H_2O$$

HOCl acts as a microbicidal agent by reacting with amines of microbes to produce chloramines:

$$HOCl + R - NH2 \rightarrow R - NHCl + H_2O$$

Another potent bactericidal compound, hydroxyl radical (OH•), is formed by the reaction of H_2O_2 and O_2^-, which requires a trace metal such as iron as catalyst (Fenton reaction):

$$\begin{array}{l} Fe^{3+} + O_2^- \rightarrow Fe^{2+} + O_2 \\ Fe^{2+} + H_2O_2 \rightarrow Fe^{3+} + OH^- + OH\bullet \\ \hline O_2^- + H_2O_2 \rightarrow O_2 + OH^- + OH\bullet \end{array}$$

Hydrogen peroxide can also be produced by the antibody-catalyzed reaction of singlet molecular oxygen ($^1O_2{*}$) and water (Wentworth et al. 2002). This reaction also produces a molecular species with a chemical structure similar to that of ozone. This species was also generated during the oxidative burst of activated human neutrophils suggesting that alternative pathways may exist for biological killing of bacteria.

In addition to ROS, the eradication of some bacteria also requires proteolytic enzymes. Neutrophil elastase, for example, has been shown to be required for efficient killing of some gram-negative bacteria (Belaaouaj et al. 1998). Both elastase and cathepsin G are required for protection against infection by *Aspergillus fumigatus* (Tkalcevic et al. 2000). In a recent study (Reeves et al. 2002), mice deficient in cathepsin G were able to resist *Candida albicans*, but not *Staphylococcus aureus*, indicating that cathepsin G is required for immunity against *S. aureus*. Mice that were deficient in elastase, on the contrary, were able to defend themselves against *S. aureus*, but not against *C. albicans*. In vitro experiments using neutrophils purified from these mice revealed that phagocytosis, degranulation, oxidase activity, and myeloperoxidase activity were normal, thus supporting a primary role for proteases in bacterial killing. Indeed, treatment of normal human neutrophils with a cocktail of protease inhibitors disenabled the ability of these neutrophils to kill *S. aureus*.

On the basis of these observations these investigators examined the mechanism by which protease activation occurs within the phagosome (Reeves et al. 2002; Ahluwahlia et al. 2004). The prevailing dogma has been that the negative charges that accumulate in the phagosome because of the electrogenic generation of O_2^- are balanced by the positive charges of protons that enter the phagosome through fusion with acid granules and via activation of proton pumps (DeCoursey 2003; Henderson 1988; Touret and Grinstein 2002). Ahluwahlia and associates observed, however, that the pH within the phagosome increased during NADPH oxidase activity. The authors state that if all of the charge compensation were dependent on protons, then the pH would not rise from 6 to 8, as observed. (The elevated pH is kept within a physiological range by the acid granules, which provide a buffering effect according to these authors.) They suggest in their recent reports that the electron charge is fully compensated for by a large influx of potassium ions into the phagosome. This K^+ influx causes the vacuole to become hypertonic, leading to the release of proteases attached to the strongly anionic, sulfated proteoglycan granule matrix, where they are normally restrained in the unstimulated cell. The elevated pH resulting from K^+ influx supports the activation of proteases that would normally be inactive at lower pH values. Surprisingly, these authors claim that H_2O_2 had no microbicidal activity. even in the presence of Cl^- and myeloperoxidase, suggesting that HOCl is not an effective antimicrobial agent. Although this hypothesis provides a novel paradigm for bacterial killing, it remains highly controversial whether a K^+ influx of the magnitude proposed by these authors can be physiologically tolerated by neutrophils.

Furthermore, the concept that charge compensation via H^+ translocation, a well-established observation, plays only a minor role in the equilibrium of the phagosome is questionable (see DeCoursey 2004; Harrison et al. 2002; Roos and Winterbourn 2002 for rebuttals of this model).

8 Phagocytosis: The Lesser Evil

In light of the ROS generated and the proteases that are active in the phagosome, phagocytosis would certainly seem to be a death sentence for many bacteria. Yet a number of bacteria manage to escape death by converting the phagosome from a death chamber into a refuge. These bacteria carry out covert operations that we are only beginning to understand that allow their survival within the phagosome (Allen 2003; Cosart and Samsonetti 2004; Rosen 2004). These include preventing fusion of the phagosome with endosomes and lysosomes, thereby cutting off supplies of bactericidal proteases and microbicidal enzymes that reside in these compartments. Some bacteria opt to retreat out of the phagosome and set up camp in the cytosol rather than combat ROS or reroute trafficking of granules and lysosomes. Other bacteria can modulate neutrophil apoptotic responses (reviewed in DeLeo 2004). Apoptotic PMN do not respond to chemoattractants, phagocytose, or undergo a respiratory burst. Thus, by causing PMN to become apoptotic prematurely, bacteria such as *Escherichia coli* and *Candida albicans* can escape destruction by PMN. By delaying apoptosis, bacteria such as *Leishmania major* (Laskay et al. 2003) ensure that their neutrophil host will be engulfed by macrophages, their final host. Clearance of apoptotic PMN by macrophages is considered a normal housekeeping duty of macrophages and therefore does not elicit an antimicrobial response.

An obvious tactic that bacteria use to survive within the phagosome is to halt ROS production by the NADPH oxidase, which we will focus on in this final section (Fig. 4). As in chess, there are many strategies that lead to a checkmate, and it is always wise to have more than one strategy in play. As mentioned in the introduction of this chapter, altering the activation state of Rac (or Cdc 42 as discussed above) by bacterially produced upstream signaling proteins (GAPs, GEFs, protein phosphatases) or toxins would certainly affect ROS production along with other functions of the host cell. Not much is known about how other components of the NADPH oxidase might be regulated by pathogenic bacteria, but recent studies on the obligate intracellular PMN pathogen *Anaplasma phagocytophilum* provide an example that establishes that regulation of other components does occur. *A. phagocytophilum* is

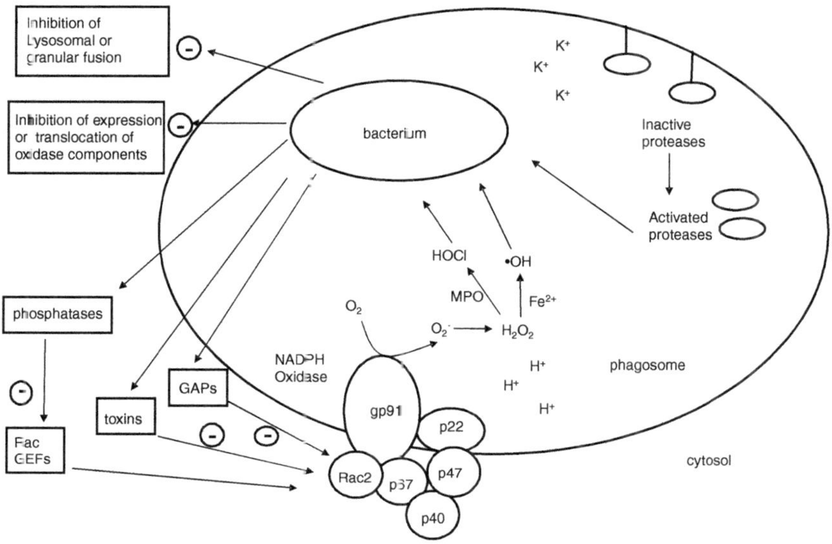

Fig. 4 Disarmament of the NADPH oxidase. There are several mechanisms by which bacteria can survive after they are engulfed by a phagocytic granulocyte. By preventing lysosomal and granular fusion with the phagosome, bacteria can cut off supplies of myeloperoxidase (*MPO*), proteases, gp91^{phox}, and p22^{phox}. By inhibiting the expression or translocation of oxidase components, bacteria can also compromise NADPH oxidase assembly. Because activation of Rac is essential for functional NADPH oxidase activity, production of GAPs, Rac GTPase-targeting toxins, or phosphatases that inhibit RacGEFs will decrease ROS production. Other possible mechanisms might include activation of Cdc42, which antagonizes Rac-induced NADPH oxidase activation, or integrin activation that would cause a delayed oxidative burst because of Rac inactivation (see text for details)

the agent that causes human granulocytic ehrlichiosis (HGE), a disease that is transmitted from ticks to humans. Phagocytosis of *A. phagocytophilum* by neutrophils does not induce an oxidative burst, and within 30 min of uptake it prevents NADPH oxidase activation by a number of stimuli in vitro, including fMLF and PMA (Mott et al. 2002). A recent study observed that p22^{phox} protein levels, but not levels of gp91^{phox}, p47^{phox}, p40^{phox}, or p67^{phox}, decreased within 30 min after exposure of human PMN to *A. phagocytophilum* (Mott et al. 2002). In granulocyte-differentiated HL-60 cells, this group also observed decreased p22^{phox} levels, but only at 7 days post infection. Because translocation of p47^{phox}, which docks on p22^{phox}, was not prevented by *A. phagocytophilum* infection, the authors of this study could not rule out the possibility that p22^{phox} was modified or degraded, with loss of the C-terminal region recognized by their antibody. In another study on *A. phagocytophilum*

(Carlyon et al. 2002), total RNA from differentiated HL-60 cells was isolated from uninfected and infected cells to generate cDNA probes that were used in a leukocyte gene array analysis. Rac 2 was among the genes that were downregulated on infection. Examination of Rac2 mRNA levels by quantitative RT-PCR revealed that Rac2 gene expression was decreased sevenfold in differentiated HL-60 cells 48 h post infection (hpi) and by 50-fold in PMN 24 hpi. Western immunoblot analysis of infected and noninfected differentiated HL-60 cells showed that infection by *A. phagocytophila* also reduced Rac2 proteins levels 48 hpi. Using similar methods, these authors had shown previously that gp91^{phox} gene expression and protein levels were decreased in infected differentiated HL-60 cells (Banerjee et al. 2000). In the most recent study, the authors showed that transfection of a plasmid bearing both gp91^{phox} and Rac1 into infected differentiated HL-60 cells restores NADPH oxidase activity and bacterial killing.

Another bacterial species, *Salmonella typhimurium*, inhibits ROS production by NADPH oxidase by preventing assembly of the NADPH oxidase complex. It has been observed that *S. typhimurium* mutants that have defects in *Salmonella* pathogenicity island-2 (SPI-2)-encoded components of the type III secretion system (TTSS) are less virulent in wild-type mice, but not in mice that lack functional NADPH oxidase activity (Vazquez-Torres et al. 2000) This suggests that inhibition of the NADPH oxidase by *S. typhimurium* is SPI dependent. In another study (Gallois et al. 2001), phagosomes were isolated from human monocyte-derived macrophages exposed to *S. typhimurium*. Intracellular ROS production was observed in only 13%–25% of phagosomes containing wild-type (WT) *S. typhimurium*, whereas 75%–85% of phagosomes containing the SPI-2 mutants sseD (secretion system potential effector) and ssaT (secretion system apparatus) were observed to produce ROS. Immunochemical staining of these macrophages revealed that although 70%–80% of sseD- and ssaT-mutant *S. typhimurium*-containing phagosomes were enriched for cyt b after 1 min of phagocytosis, only 40% of the WT *S. typhimurium*-containing phagosomes were positive for cyt b. By 20 min, only 25% of the WT-containing phagosomes were cyt b-positive, in contrast to 80% for the mutant-containing phagosomes. Moreover, p47^{phox} and p67^{phox} localized with cyt b in the mutant *S. typhimurium*-containing phagosomes but were not detected in the phagosomal membranes of cyt b-negative phagosomes containing WT *S. typhimurium*.

Listeria monocytogenes is an intracellular facultative bacterium that serves as an example of a bacterium that escapes the phagosome and thrives in the cytoplasm of nonstimulated macrophages (reviewed in Vazquez-Boland et al 2001). *L. monocytogenes* secretes two proteins, listeriolysin and phosphatidylinositol-phospholipase C, which lyse the phagosomal membrane and allow *L. monocytogenes* to escape into the cytosol. *L. monocytogenes* infection does not activate ROS production in macrophages. While in the phagosome, it inactivates Rab5a, a regulator of endosomal trafficking (Prada-Delgado 2001). This in turn prevents lysosomal fusion and prevents lysosomal proteases LAMP-1 and cathepsin D from entering the phagosome. It was shown, however, that clearance of *L. monocytogenes* is enhanced by interferon-gamma (IFN-γ) stimulation of macrophages. ROS generated by NADPH oxidase and reactive nitrogen intermediates produced by nitric oxide synthase are important mediators of *L. monocytogenes* killing (Myers et al 2003). IFN-γ actually enhances NADPH oxidase activity though the Rab5-induced remodeling of the phagosomal membrane, facilitating the association of Rac2 with *L. monocytogenes*-containing phagosomes, thereby increasing ROS formation via NADPH oxidase (Prado-Delgado 2001).

References

Abo A, Pick E, Hall A, Totty N, Teahan CG, Segal AW (1991) Activation of the NADPH oxidase involves the small GTP-binding protein p21^{rac1}. Nature 353:668–670

Ahluwalia J, Tinker A, Clapp LH, Duchen MR, Abramov AY, Pope S, Nobles M, Segal AW (2004) The large-conductance Ca^{2+}-activated K^{+} channel is essential for innate immunity. Nature 427:853–858

Allen LA (2003) Mechanisms of pathogenesis: evasion of killing by polymorphonuclear leukocytes. Microbes Infect 5:1329–1335

Aghazadeh B, Lowry WE, Huang XY, Rosen MK (2000) Structural basis for relief of autoinhibition of the Dbl homology domain of proto-oncogene Vav by tyrosine phosphorylation. Cell 102:625–633

Babior BM (1999) NADPH oxidase: An update. Blood 93:1464–1476

Babior BM, Lambeth JD, Nauseef W (2002) The neutrophil NADPH oxidase. Arch Biochem Biophys 397:342–344

Banerjee R, Anguita J, Roos D, Fikrig E (2000) Cutting edge: infection by the agent of human granulocytic ehrlichiosis prevents the respiratory burst by down-regulating gp91phox. J Immunol 164:3946–3949

Barbieri JT, Riese MJ, Aktories K (2002) Bacterial toxins that modify the actin cytoskeleton. Annu Rev Cell Dev Biol 18:315–344

Belaaouaj A, McCarthy R, Baumann M, Gao Z, Ley TJ, Abraham SN, Shapiro SD (1998) Mice lacking neutrophil elastase reveal impaired host defense against Gram-negative bacterial sepsis. Nat Med 4:614–618

Benard V, Bohl BP, Bokoch GM (1999) Characterization of Rac and Cdc42 activation in chemoattractant-stimulated human neutrophils using a novel assay for active GTPases. J Biol Chem 274:13198–13204

Bishop AL, Hall A (2000) Rho GTPases and their effector proteins. Biochem J 348:241–255

Bliska JB (2000) Yop effectors of *Yersinia* spp and actin rearrangements. Trends Microbiol 8:205–208

Bokoch GM and Diebold BA (2002) Current molecular models for NADPH oxidase regulation by Rac GTPase. Blood 100:2692–2696

Bokoch GM, Bohl BP, Chuang TH (1994) Guanine nucleotide exchange regulates membrane translocation of Rac/RhoGTP-binding proteins. J Biol Chem 269:31674–31679

Bokoch GM, Knaus UG (2003) NADPH oxidases: not just for leukocytes anymore! Trends Biochem Sci 28:502–508

Boyle EC, Finlay BB (2003) Bacterial pathogenesis: exploiting cellular adherence. Curr Opin Cell Biol 15:633–639

Burridge K, Wennerberg K (2004) Rho and Rac take center stage. Cell 116:167–179

Carlyon JA, Chan WT, Galan J, Roos D, Fikrig E (2002) Repression of rac2 mRNA expression by *Anaplasma phagocytophila* is essential to the inhibition of superoxide production and bacterial proliferation. J Immunol 169:7009–7018

Chuang TH, Bohl BP, Bokoch GM (1993) Biologically active lipids are regulators of Rac-GDI complexation. J Biol Chem 268:26206–26211

Cornelis GR (2002) The *Yersinia* YSC-YOP "Type III" weaponry. Nat Rev Mol Cell Biol 3:742–752

Cossart P, Sansonetti PJ (2004) Bacterial invasion: The paradigms of enteroinvasive pathogens. Science 304:242–248

DeCoursey TE (2004) During the respiratory burst, do phagocytes need proton channels or potassium channels or both? Sci STKE pp.pe21

DeCoursey TE, Morgan D, Cherny VV (2003) The voltage dependence of NADPH oxidase reveals why phagocytes need proton channels. Nature 422:531–534

DeLeo FR (2004) Modulation of phagocyte apoptosis by bacterial pathogens. Apoptosis 9:399–413

DerMardirossian C, Schnelzer A, Bokoch GM (2004) Phosphorylation of RhoGDI by Pak1 mediates dissociation of Rac GTPase. Mol Cell 15:117–127

Dharmawardhane S, Bokoch GM (1997) Rho GTPases and leukocyte cytoskeletal regulation. Curr Opin Hematol 4:12–18

Diebold BA, Bokoch GM (2001) Molecular basis for Rac2 regulation of phagocyte NADPH oxidase. Nat Immunol 2:211–215

Diebold BA, Fowler B, Lu J, Dinauer MC, Bokoch GM (2004) Antagonistic crosstalk between Rac and Cdc42 GTPases regulates generation of reactive oxygen species. J Biol Chem 279:28136–28142

Dinauer MC (2003) Regulation of neutrophil function by Rac GTPases. Curr Opin Hematol 10:8–15

Dorseuil O, Vazquez A, Lang P, Bertoglio J, Gacon G, Leca G (1992) Inhibition of superoxide production in B lymphocytes by rac antisense oligonucleotides. J Biol Chem 267:20540–20542

Dorseuil O, Quinn MT, Bokoch GM (1995) Dissociation of Rac translocation from p47/p67 movements in human neutrophils by tyrosine kinase inhibitors. J. Leukoc Biol 58:108–113

Ellson CD, Gobert-Gosse S, Anderson KE, Davidson K, Erdjument-Bromage H, Tempst P, Thuring JW, Cooper MA, Lim ZY, Holmes AB, Gaffney PR, Coadwell J, Chilvers ER, Hawkins PT, Stephens LR (2001) PtdIns(3)P regulates the neutrophil oxidase complex by binding to the PX domain of p40(phox). Nat Cell Biol 3:679–682

Freeman JL, Lambeth (1996) NADPH oxidase activity is independent of $p47^{phox}$ in vitro. J Biol Chem 271:22578–22582

Gallois A, Klein JR, Allen LA, Jones BD, Nauseef WM (2001) *Salmonella* pathogenicity island 2-encoded type III secretion system mediates exclusion of NADPH oxidase assembly from the phagosomal membrane. J Immunol 166:5741–5748

Groemping Y, Lapouge K, Smerdon SJ, Rittinger K (2003) Molecular basis of phosphorylation-induced activation of the NADPH oxidase. Cell 113:343–355

Han CH, Freeman JL, Lee T, Motalebi SA, Lambeth JD (1998) Regulation of the neutrophil respiratory burst oxidase: identification of an activation domain in p67^{phox}. J Biol Chem 273:16663–16668

Harrison RE, Touret N, Grinstein S (2002) Microbial killing: oxidants, proteases and ions. Curr Biol 12:R357-R359

Henderson LM, Chappell JB, Jones OTG (1988) Internal pH changes associated with the activity of NADPH oxidase of human neutrophils: Further evidence for the presence of an H^+ conducting channel. Biochem J 251:563–567

Heyworth PG, Bohl BP, Bokoch GM, Curnutte JT (1994) Rac translocates independently of the neutrophil NADPH oxidase components p47^{phox} and p67^{phox}: Evidence for its interaction with flavocytochrome b558. J Biol Chem 269:30749–30752

Heyworth PG, Cross AR, Curnutte JT (2003) Chronic granulomatous disease. Curr Opin Immunol 15:578–584

Hoppe AD, Swanson JA (2004) Cdc42, Rac1, and Rac2 display distinct patterns of activation during phagocytosis. Mol Biol Cell 15:3509–2519

Kanai F, Liu H, Field SJ, Akbary H, Matsuo T, Brown GE, Cantley LC, Yaffe MB (2001) The PX domains of p47^{phox} and p40 phox bind to lipid products of PI(3)K Nat Cell Biol 3:675–678

Knaus UG, Heyworth PG, Evans T, Curnutte JT, Bokoch GM (1991) Regulation of phagocyte oxygen radical production by the GTP-binding protein Rac2. Science 254:1512–1515

Koshkin V, Lotan O, Pick E (1996) The cytosolic component p47^{phox} is not a *sine qua non* participant in the activation of NADPH oxidase but is required for optimal superoxide production. J Biol Chem 271:30326–30329

Kwong, CH, Adams AG, Leto TL (1995) Characterization of the effector-specifying domain of Rac involved in NADPH oxidase activation. J Biol Chem 270:19868–19872

Lapouge K, Smith SJ, Walker PA, Gamblin SJ, Smerdon SJ, Rittinger K (2000) Structure of the TPR domain of p67^{phox} in complex with Rac GTP. Mol Cell 6:899–907

Lapouge K, Smith SJ, Groemping Y, Rittinger K (2002) Architecture of the p40-p47-p67^{phox} complex in the resting state of the NADPH oxidase. A central role for p67^{phox}. J Biol Chem 277:10121–10128

Laskay T, van Zandbergen G, Solbach W(2003) Neutrophil granulocytes—Trojan horses for *Leishmania major* and other intracellular microbes? Trends Microbiol 11:210–214

Li Z, Hannigan M, Mo Z, Liu B, Lu W, Wu Y, Smrcka AV, Wu G, Li L, Liu M, Huang CK, Wu D (2003) Directional sensing requires G$\beta\gamma$ mediated Pak 1 and Pixαdependent activation of Cdc42. Cell 114:215–227

Matozaki T, Nakanishi H, Takai Y (2000) Small G-protein networks: their crosstalk and signal cascades. Cell Signal 12:515–524

Moores SL, Selfors LM, Fredericks J, Breit T, Fujikawa K, Alt FW, Brugge JS, Swat W (2000) Vav family proteins couple to diverse cell surface receptors. Mol Cell Biol 20:6364–6373

Mott, J, Rikihisa, Y, Tsunawaki S (2002) Effects of *Anaplasma phagocytophila* on NADPH oxidase components in human neutrophils and HL-60 cells. Infect Immun 70:1359–1366

Myers JT, Tsang AW, Swanson JA (2003) Localized reactive oxygen and nitrogen intermediates inhibit escape of *Listeria monocytogenes* from vacuoles in activated macrophages. J Immunol 171:5447–5453

Nisimoto Y, Motalebi S, Han CH, Lambeth JD (1999) The $p67^{phox}$ activation domain regulates electron flow from NADPH to flavin in flavocytochrome b_{558}. J Biol Chem 274:22999–23005

Prada-Delgado A, Carrasco-Marin E, Bokoch GM, Alvarez-Dominguez C (2001) Interferon-γ listericidal action is mediated by novel Rab5a functions at the phagosomal environment. J Biol Chem 276:19059–19065

Price MO, McPhail LC Lambeth JD Han CH, Knaus UG, Dinauer MC (2002) Creation of a genetic system for analysis of the phagocyte respiratory burst high-level reconstitution of the NADPH oxidase in a nonhematopoietic system. Blood 99:2653–2661

Quinn MT, Evans T, Loetterle LR, Jesaitis AJ and Bokoch GM (1993) Translocation of Rac correlates with NADPH oxidase activation. J Biol Chem 268:20983–20987

Reeves EP, Lu H, Jacobs HL, Messina CG, Bolsover S, Gabella G, Potma EO, Warley A, Roes J, Segal AW (2002) Killing activity of neutrophils is mediated through activation of proteases by K^+ flux. Nature 416:291–297

Roberts AW, Kim C, Zhen L, Lowe JB, Kapur R, Petryniak B, Spaetti A, Pollock JD, Borneo JB, Bradford GB, Atkinson SJ, Dinauer MC, Williams DA (1999) Deficiency of the hematopoietic cell-specific Rho family GTPase Rac2 is characterized by abnormalities in neutrophil function and host defense. Immunity 10:183–196

Roos D, Winterbourn CC (2002) Lethal weapons. Science 296:669–671

Roos D, van Bruggen R, Meischl C (2003) Oxidative killing of microbes by neutrophils. Microbes Infect 5:1307–1315

Rosen H (2004) Bacterial responses to neutrophil phagocytosis. Curr Opin Hematol 11:1–6

Sibley, LD (2004) Intracellular parasite invasion strategies. Science 304:248–253

Srinivasan S., Wang F, Glavas S, Ott A, Hofmann F, Aktories K, Kalman D, Bourne HR (2003) Rac and Cdc42 play distinct roles in regulating PI(3,4,5)P3 and polarity during neutrophil chemotaxis. J Cell Biol 160:375–385

Stebbins CE, Galan JE (2001) Structural mimicry in bacterial virulence. Nature 412:701–705

Touret N, Grinstein S (2002) Voltage-gated proton "channels": a spectator's viewpoint. J Gen Physiol 120:767–771

Tkalcevic J, Novelli M, Phylactides M, Iredale JP, Segal AW, Roes J (2000) Impaired immunity and enhanced resistance to endotoxin in the absence of neutrophil elastase and cathepsin G. Immunity 12:201–210

Vazquez-Boland JA, Kuhn M, Berche P, Chakraborty T, Dominguez-Bernal G, Goebel W, Gonzalez-Zorn B, Wehland J, Kreft J(2001) Listeria pathogenesis and molecular virulence determinants. Clin Microbio Rev 14:584–640

Vazquez-Torres A, Xu Y, Jones-Carson J, Holden DW, Lucia SM, Dinauer MC, Mastroeni P, Fang FC (2000) *Salmonella* pathogenicity island 2-dependent evasion of the phagocyte NADPH oxidase. Science 287:1655–1658

Voncken JW, van Schaick H, Kaartinen V, Deemer K, Coates T, Landing B, Pattengale P, Dorseuil O, Bokoch GM, Groffen J, Heisterkamp N (1995) Increased neutrophil respiratory burst in bcr-null mutants. Cell 80:719–728

Wentworth, Jr P, McDunn JE, Wentworth AD, Takeuchi C, Nieva J, Jones T, Bautista C, Ruedi JM, Gutierrez A, Janda KD, Babior BM, Eschenmoser A, Lerner RA (2002) Evidence for antibody-catalyzed ozone formation in bacterial killing and inflammation. Science 298:2195–2199

Wientjes FB, Segal AW (2003) PX domain takes shape. Curr Opin Hematol 10:2–7

Yuzawa S, Suzuki NN, Fujioka Y, Ogura K, Sumimoto H, Inagaki F (2004) A molecular mechanism for autoinhibition of the tandem SH3 domains of p47^{phox}, the regulatory subunit of the phagocyte NADPH oxidase. Genes Cells 9:443–456

Zhao T, Benard V, Bohl BP, Bokoch GM (2003) The molecular basis for adhesion-mediated suppression of reactive oxygen species generation by human neutrophils. J Clin Invest 112:1732–1740

Zhao X, Carnevale KA, Cathcart MK (2003) Human monocytes use Rac1, not Rac2, in the NADPH oxidase complex. J Biol Chem 278:40788–40792

CTMI (2005) 291:113–145

Clostridial Rho-Inhibiting Protein Toxins

K. Aktories[1] (✉) · I. Just[2]

[1]Institut für Experimentelle und Klinische Pharmakologie und Toxikologie, Albertstrasse 25, 79104 Freiburg, Germany
klaus.aktories@pharmakol.uni-freiburg.de

[2]Institut für Toxikologie, Medizinische Hochschule Hannover, Carl-Neuberg-Str. 1, 30625 Hannover, Germany

Abstract Rho proteins are master regulators of a large array of cellular functions, including control of cell morphology, cell migration and polarity, transcriptional activation, and cell cycle progression. They are the eukaryotic targets of various bacterial protein toxins and effectors, which activate or inactivate the GTPases. Here Rho-inactivating toxins and effectors are reviewed, including the families of large clostridial cytotoxins and C3-like transferases, which inactivate Rho GTPases by glucosylation and ADP-ribosylation, respectively.

1
Rho Proteins as Targets of Toxins

Rho GTPases (Ras GTPases for some toxins) are the predominant cellular targets of clostridial glucosylating toxins and C3-like ADP-ribosylating exoenzymes. These GTPases act as molecular switches in various signaling pathways (Bishop and Hall 2000; Etienne-Manneville and Hall 2002; Takai et al. 2001; Wennerberg and Der 2004). The switch proteins are regulated by a GTPase cycle, which is described in great detail elsewhere in this volume. Rho proteins are inactive in the GDP-bound form and localized in the cytosol. In the active, signaling GTP-bound form, the isoprenylated proteins are at the plasma membranes. In the cytosol, the Rho proteins are complexed with the guanine nucleotide dissociation inhibitor (GDI), which keeps them in the cytosol and prevents nucleotide exchange. Membrane receptor-mediated activation triggers translocation and dissociation of the Rho-GDI complex. The nucleotide exchange of GDP to GTP is catalyzed by guanine exchange factors (>60 GEFs), resulting in the active GTP-bound form of the GTPases.

Fig. 1A–C. Molecular mode of action of toxin A/B. **A** The Rho-GTPases are molecular switches that are regulated by guanine nucleotide binding. Nonsignaling, inactive Rho is complexed with the guanine nucleotide dissociation inhibitor GDI residing in the cytosol. Signal input induces an activation cascade resulting in translocation to the plasma membranes and nucleotide exchange catalyzed by guanine nucleotide exchange factor (*GEF*). GTP binding causes a conformational change of especially the effector region, allowing Rho to interact with effector proteins. The effector proteins comprise Thr/Ser-kinases, lipid kinases, lipases, or scaffold proteins that execute and amplify Rho signals. Rho signaling is terminated by an additional regulatory protein called GTPase-activating protein (*GAP*) that increases GTP hydrolysis, resulting in inactive GDP-bound Rho, which is delivered to the cytosol again. Activation and inactivation of Rho-GTPases are governed by GTPase- and cytosol-membrane cycling. **B** Toxin-catalyzed monoglucosylation alters the properties of Rho-GTPases. Glucosylation leads to an entrapment of Rho-GTPases at membranous binding sites (p70) and prevents activation by GEF (*1*). The glucose moiety stabilizes the effector region so that Rho-GTPases are incapable of interacting with their effector proteins, thereby completely blocking downstream signaling (*2*). Glucosylated Rho-GTPases are insensitive toward GAP (*3*). Furthermore, glucosylation inhibits binding to GDI, thereby preventing extraction of Rho from the membranes (*4*). Entrapment at restricted membranous binding sites and interruption of cytosol-membrane cycling is the basis of the complete inhibition of all Rho-dependent signal pathways. **C** C3-catalyzed ADP-ribosylation of RhoA does not block nucleotide binding/exchange and does not prevent effector coupling. However, binding to GDI is increased so that ADP-ribosylated Rho is entrapped in the cytosolic GDI complex (*2*). Release of RhoA seems to be impossible, resulting in an interruption of the cytosol-membrane cycling of Rho. Furthermore, activation by GEF is inhibited (*1*) ▶

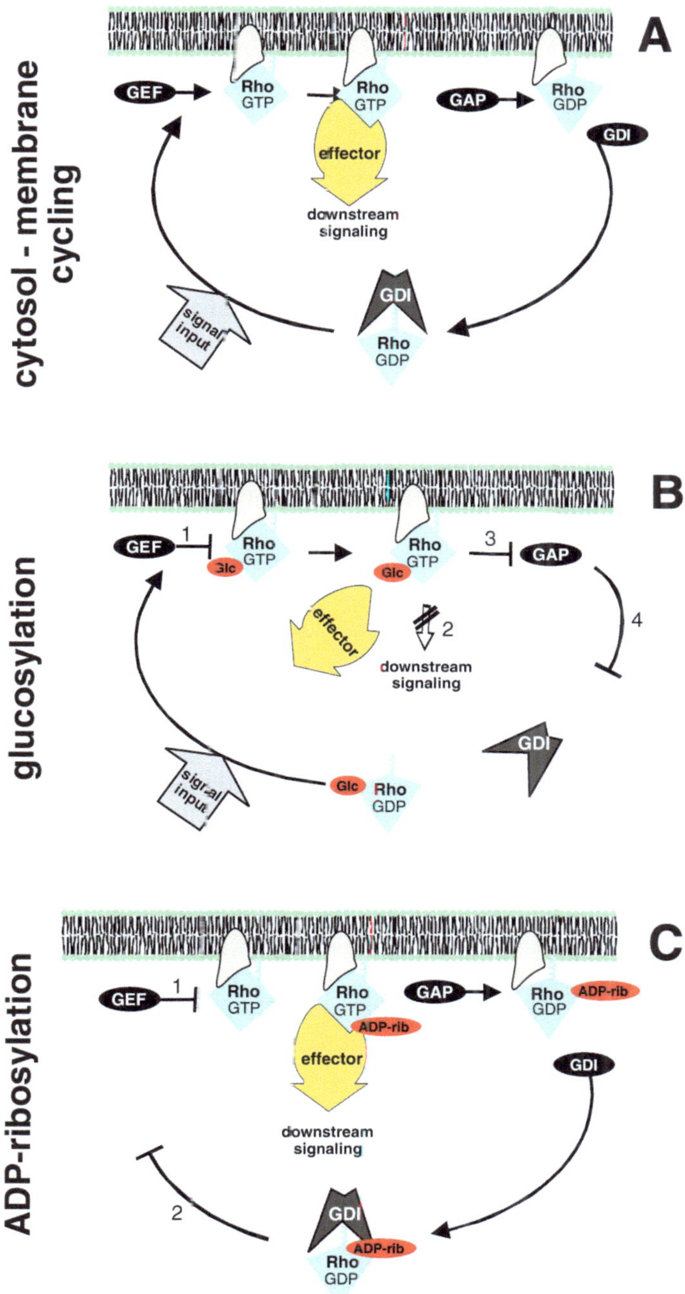
A
cytosol - membrane cycling
GEF
Rho GTP
Rho GTP
GAP
Rho GDP
GDI
effector
downstream signaling
signal input
GDI
Rho GDP
B
glucosylation
GEF
1
Glc
Rho GTP
Glc
Rho GTP
3
GAP
effector
2
downstream signaling
4
GDI
signal input
Glc
Rho GDP
C
ADP-ribosylation
GEF
1
Rho GTP
Rho GTP
ADP-rib
GAP
Rho GDP
ADP-rib
effector
GDI
downstream signaling
2
GDI
ADP-rib
Rho GDP

GTP binding results in a conformational change of switch 1, which allows binding to and activation of downstream effectors. Proteins of the family of GTPase-activating proteins (>60 GAPs) cause accelerated hydrolysis of GTP to GDP and switch off the active state of small GTPases (see Fig. 1A).

The family of Rho GTPases comprise Rho (A, B, C, D,G), Rac (1, 1b, 2, 3), Cdc42, Wnt-1, Chp, G25 K, Rnd (1,2,3), TTF/RhoH, Rif, and TC10 (for review see Mackay and Hall 1998; Ridley 2000; Van Aelst and D'Souza-Schorey 1997). Best studied are the RhoA, Rac, and Cdc42 subtypes. Whereas RhoA induces formation of actin stress fibers and focal adhesions (Etienne-Manneville and Hall 2002), Rac leads to formation of lamellipodia and membrane ruffles (Ridley et al. 1992) and Cdc42 induces formation of microspikes/filopodia (Kozma et al. 1995). Thus Rho GTPases are essential for cell migration, control of morphogenesis, and cell polarity. Multiple additional regulatory functions of Rho GTPase have been shown, including cell cycle control, activation of transcription, apoptosis, and transformation (for review see Bishop and Hall 2000; Etienne-Manneville and Hall 2002; Takai et al. 2001; Wennerberg and Der 2004)).

Rho GTPases are the targets of various bacterial protein toxins, which cause either activation or inactivation of the target GTPase. Activation of Rho GTPases is achieved by deamidation mediated by cytotoxic necrotizing factors CNF1, CNF2, and CNFY produced by *Escherichia coli* and *Yersinia pseudotuberculosis*, respectively (Flatau et al. 1997; Hoffmann et al. 2004; Schmidt et al. 1997). The related transglutamination caused by the dermonecrotic toxin from *Bordetella* species also activates Rho GTPases (Masuda et al. 2000). Rho GTPases are inhibited by glucosylation and ADP-ribosylation catalyzed by the family of clostridial glucosylating toxins and C3-like exoenzymes, respectively. Inhibition of Rho GTPases is also caused by bacterial effectors translocated into the target cells by the type III secretion system. These include the *Yersinia* protease YopT (Juris et al. 2002; Shao et al. 2002), which cleaves the very C-terminal part of Rho, and various bacterial GAPs (YopE from *Yersinia*, ExoS from *Pseudomonas aeruginosa*, and SptP from *Salmonella*), which decrease the level of active GTP-bound forms of Rho GTPases by mimicking the turn-off function of eukaryotic GTPase-activating proteins (GAPs) (Barbieri et al. 2002; Fu and Galán 1999; Goehring et al. 1999; von Pawel-Rammingen et al. 2000). Here we describe the structure-function analysis, biological effects, and functional consequences of glucosylation and ADP-ribosylation catalyzed by clostridial bacterial protein toxins.

2
Clostridial Glucosylating Toxins

Members of the family of large clostridial cytotoxins are toxin A and B from *Clostridium difficile*, the hemorrhagic and lethal toxin from *Clostridium sordellii*, and the α-toxin from *Clostridium novyi*. These toxins share sequence identities ranging from 36% to 90% and have molecular masses between 250 and 308 kDa (Busch and Aktories 2000; Von Eichel-Streiber et al. 1996). Recently, several toxin isoforms have been described that considerably extend this toxin family (Table 1).

Table 1 Protein substrates and cosubstrates of the clostridial glucosylating toxins

Toxin	Molecular mass	Sugar donor	Transferred moiety	Protein substrates
C. difficile toxin A-10463	308 kDa	UDP-glucose	Glucose	Rho, Rac, Cdc42, RhoG, TC10 (Rap)
C. difficile toxin B-10463	270 kDa	UDP-glucose	Glucose	Rho, Rac, Cdc42, RhoG, TC10
C. difficile toxin B-1470	269 kDa	UDP-glucose	Glucose	Rac, R-Ras, Ral, Rap
C. difficile toxin B-8864	269 kDa	UDP-glucose	Glucose	Rac, (Cdc42), R-Ras, Ral, Rap
C. difficile toxin B-C34	269 kDa	UDP-glucose	Glucose	Rho, Rac (Cdc42), R-Ras, Ral, Rap
C. sordellii lethal toxin-6018	271 kDa	UDP-glucose	Glucose	Rac, (Cdc42), Ha-Ras, R-Ras, Ral, Rap
C. sordellii hemorrhagic toxin-9048	∼300 kDa	UDP-glucose	Glucose	Rho, Rac, Cdc42
C. novyi α-toxin-19402	250 kDa	UDP-N-acetyl-glucosamine (UDP-glucose)	N-acetyl-glucosamine (Glucose)	Rho, Rac, Cdc42

Poor substrates and/or substrates only detected in vitro are set in parentheses.
Numbers behind toxin name indicate isoform the producing strain.
The accession numbers are available at http://afmb.cnrs-mrs.fr/CAZY/GT_44.html.

2.1
Clostridium difficile Toxins A and B

Clostridium difficile toxins A and B are the major causative factors of the antibiotic-associated diarrhea and pseudomembranous colitis that result when normal intestinal flora is altered during therapy with broad-spectrum antibiotics. Infection with and subsequent overgrowth with *Clostridium difficile* with subsequent toxin production lead to inflammation and damage of the colonic mucosa (Bartlett 2002; Kelly and Lamont 1998). Major progress in this field was possible when *Clostridium difficile* was recognized as the causative agent of pseudomembranous colitis and not, as earlier suggested, *Staphylococcus aureus* or viruses (Bartlett et al. 1978; Larson et al. 1978; Larson and Proce 1977). Very important was also the observation that *Clostridium sordellii* antitoxin neutralized the cytotoxic activity of a toxin isolated from patients with colitis (Rifkin et al. 1977). Subsequently, it was found that at least two toxins are produced by *Clostridium difficile* (Taylor et al. 1981). In the 1990s, major milestones in toxin research were the cloning and sequencing of the toxin coding genes (Barroso et al. 1990; Dove et al. 1990; Sauerborn and Von Eichel-Streiber 1990; Von Eichel-Streiber et al. 1992).

In addition to their in vivo effects to cause diarrhea and colitis, both toxins are cytotoxic at cultivated cells. Toxin B was found to be 100- to 1,000-fold more toxic than toxin A in almost all cultured cells; therefore it was designated "*cytotoxin*." In many animal models, however, enterotoxicity was only associated with toxin A, which was therefore named *enterotoxin* (Lyerly et al. 1988; Wilkins 1987). Now it is accepted that toxin A-negative/toxin B-positive strains can also induce pseudomembranous colitis. Moreover, recently enterotoxic properties were reported for toxin B as well (Riegler et al. 1995; Savidge et al. 2003). The human colonocytes are sensitive to both toxins A and B, whereas the animal gut epithelium is almost resistant to toxin B. This discrepancy explains why toxin A was thought of for a long time as the true enterotoxin. Perhaps the most important progress in the understanding of the role of the toxins in disease was made by the findings that *Clostridium difficile* toxins A and B are glucosyltransferases, which transfer the glucose moiety from UDP-glucose to members of the Rho family of small GTPases (Just et al. 1995b, 1995c). These findings established a novel family of bacterial protein toxins, which were initially termed "large clostridial cytotoxins," because of their high molecular mass and their cytotoxic activity toward cell lines. Actually, we would suggest a novel name, *clostridial glucosylating toxin* (CGT), because we think this is a more appropriate characterization of the toxins.

2.2 Structure and Activity of Clostridial Glucosylating Toxins

With molecular masses of 250 to 308 kDa, the clostridial glucosylating toxins are some of the largest bacterial protein toxins. The single-chain toxins appear to be typical AB toxins with an enzyme domain and a binding/translocation

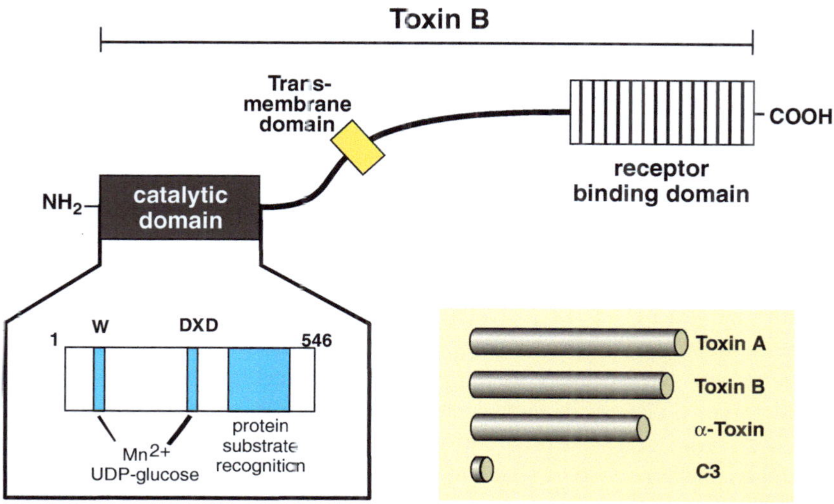

Fig. 2 Structure of the clostridial glucosylating toxins illustrated for toxin B. Toxin B includes three functional domains. *I*: The receptor-binding domain is composed of repetitive oligopeptide elements commonly accepted as the motif for binding to carbohydrate structures of the receptor. *II*: A hydrophobic region in the middle part of the molecule is supposed to form a transmembrane domain, allowing the catalytic domain to translocate into the cytoplasm. *III*: The catalytic domain residing in the N-terminal part possesses monoglucosyltransferase activity to modify the Rho-GTPases. The first 546 residues of the N-terminus are the minimum size. The tryptophan-102 (W) and the DXD motif (residues 286–288) are involved in UDP-glucose cosubstrate binding through Mn^{2+} or Mg^{2+}. The C-terminal part (residues 408–468) of the catalytic domain covers the protein substrate recognition site. The insert gives the relation of the molecular size of clostridial glucosylating toxins and C3

domain. It is now generally accepted that the catalytic domain and the receptor binding domain are located at the N-terminus and C-terminus, respectively. Based only on secondary structure prediction, the hydrophobic region in the middle part of the toxin molecule is proposed to be involved in translocation of the proteins across cellular membranes. All members of the family share the same three-domain structure (Fig. 2).

2.3 The N-Terminal Glucosyltransferase Domain

It has been convincingly shown that the glucosyltransferase activity of the toxins is N-terminally located (Busch et al. 2000b; Hofmann et al. 1997). For example, microinjection of recombinant C-terminal deletions, constructed from various members of this toxin family, caused typical morphological and cytoskeletal changes in eukaryotic target cells identical with those observed with holotoxins. In vitro glucosyltransferase activities of the C-terminally deleted toxins were very similar to native toxins. It was shown for *Clostridium difficile* toxin B and also for *Clostridium sordellii* lethal toxin that amino acids 1–546 were sufficient for full enzyme activity. By contrast, toxin truncations of the first 516 amino acids were enzymatically inactive (Hofmann et al. 1997).

Clostridium difficile toxins A and B and the variant toxin B as well as lethal and hemorrhagic toxins from *Clostridium sordellii* recruit the nucleotide sugar UDP-glucose as cosubstrate; the glucose moiety is transferred to the protein substrate (Table 1). *Clostridium novyi* α-toxin is an exception, because it uses UDP-GlcNAc as a cosubstrate. *N*-acetylglucosaminylation of small target GTPases, in fact, occurs in intact cells, as it was detected by specific [^{14}C]galactosylation of the cellular GTPase, which depends on the *N*-acetylglucosamine moiety attached (Selzer et al. 1996). α-Toxin also utilizes UDP-glucose, but the K_m is about 340 μM compared to 17 μM for UDP-glucNAc (Busch et al. 2000b). The K_m of the large glucosylating toxins for nucleotide sugars is in the range of 10–20 μM.

The clostridial glucosylating toxins differ in their substrate specificity (Table 1). Toxins A and B, the *Clostridium sordellii* hemorrhagic toxin, and the α-toxin from *Clostridium novyi* selectively glucosylate Rho subfamily GTPases but not GTPases from other subfamilies. However, *Clostridium sordellii* lethal toxin possesses a different substrate specificity: It modifies only Rac but not Rho and, in addition, H-Ras, Rap, Ral, and R-Ras (Genth et al. 1996; Hofmann et al. 1996; Just et al. 1996; Popoff et al. 1996). This special substrate specificity is not restricted to lethal toxin. Variant forms of toxin B from *Clostridium difficile* (e.g., toxin B-1470) resemble lethal toxin.

Structure-function analysis of the active fragments of *Clostridium difficile* toxin B, *Clostridium sordellii* lethal toxin, and *Clostridium novyi* α-toxin suggested that the protein substrate recognition site is located in the C-terminal region of the catalytic domain (Hofmann et al 1998). By constructing chimeras of toxins with different substrate specificities, regions in the catalytic domain were identified that are responsible for substrate recognition. These studies indicate that the protein substrate recognition by *Clostridium sordellii* lethal toxin occurs between amino acids 365 and 516 (Hofmann et al. 1998). Moreover, the recognition site is modularly organized so that in lethal toxin Rho is recognized by a different site than Ras. The region, which is involved in binding of activated sugar nucleotides, is less defined. Chimeras of *Clostridium sordellii* lethal toxin and *Clostridium novyi* α-toxin, which uses UDP-glucose and UDP-*N*-acetylglucosamine, respectively, indicate that the nucleotide-sugar interaction is between amino acids 133 and 517 of α-toxin (Busch et al. 2000b).

Recently, amino acid residues conserved among the family of clostridial glucosylating toxins were shown to be essential for enzyme activity. Of special importance is the DXD motif, which resides almost in the middle of the catalytic domain (Fig. 2). Exchange of one of the aspartates to alanine or asparagine inhibits glucosyltransferase activity (Busch et al. 1998). Many different families of glycosyltransferases possess this highly conserved DXD motif (Breton and Imberty 1999; Wiggins and Munro 1998). Although its exact role in the catalytic reaction is not well defined, it appears that in the clostridial glucosylating toxins the DXD motif is involved in nucleotide-sugar binding via manganese ions (Busch et al. 2000a). This notion is supported by crystallographic data obtained from two glycosyltransferases, β-1,3-glucuronyltransferase and α-1,4-N-acetyl-hexosaminyltransferase, which show that the DXD motif is involved in the coordination of Mn^{2+}, through which the nucleotide sugar interacts with the enzyme (Negishi et al. 2003).

In the glucosylating toxins, the DXD motif is positioned in a region of high sequence homology (Busch et al. 1998). Many different prokaryotic and eukaryotic glycosyltransferases share this "extended" DXD motif (Keusch et al. 2000). Glycosyltransferases sharing the extended DXD motif catalyze the transfer of sugar under retention of the α-configuration. This is also true for the clostridial glucosylating toxins (e.g., *Clostridium sordellii* lethal toxin), as shown by crystal structure analysis of the glucosylated Ras and NMR spectroscopy (Geyer et al. 2003; Vetter et al. 2000). Recently, the family of clostridial glucosylating toxins has been compiled as an autonomous subfamily designated "glycosyltransferase family 44" (GT44) by Henrissat et al. (available on the Internet at http://afmb.cnrs-mrs.fr/CAZY/GT_44.html).

In addition to transferase activity the toxins exhibit glycohydrolase activity, i.e., hydrolytic cleavage of nucleotide sugar in the absence of the protein substrates. Glycohydrolase activity is much slower than transferase activity, and the biological relevance is not clear. Nevertheless, glycohydrolase activity is an excellent model to study the dependence of the enzyme on divalent cations, because all interferences with the cation binding substrate GTPases are absent. In addition to divalent cations such as Mn^{2+} or Mg^{2+}, the monovalent K^+ but not Na^+ is essential for enzyme activity (Ciesla and Bobak 1998; Just et al. 1996).

2.4 Binding and Translocation Domains

The putative receptor-binding domain, which is located at the C-terminus, is characterized by small repetitive sequence motifs, also called CROPS, clostridial repetitive oligopeptides (Fig. 2). They consist of 20 to 50 amino acids and are repeated 14–30 times (Dove et al. 1990; Von Eichel-Streiber et al. 1992, 1996). Recombinant fragments of *Clostridium difficile* toxins, covering the putative binding domains, and an antibody directed against the C-terminus of *Clostridium difficile* toxin A inhibit intoxication by the holotoxins (Frey and Wilkins 1992; Sauerborn et al. 1997). However, deletion of the C-terminal repetitive domain of toxin B decreases the cytotoxicity only by a factor of 10; this finding is not in line with the hypothesis that only the CROPS define the binding site of the toxins (Barroso et al. 1994). Moreover, it was shown recently that the whole repetitive region of toxin A is needed for binding and endocytosis (Frisch et al. 2003).

2.5 Toxin Receptors

The precise nature of the membrane receptors for clostridial glucosylating toxins is not known. In many receptor-binding studies toxin A was used. From these studies and from the proposed properties of CROPs, it was suggested that toxin A binds like a lectin to Galα-1-3Galβ1-4GlcN structures (Krivan et al. 1986; Tucker and Wilkins 1991). A 160-kDa galactose- and *N*-acetylglucosamine-containing glycoprotein was purified from brush border cells of small intestine of infant hamsters, which was suggested to be a toxin receptor (Rolfe and Song 1993). Furthermore, binding of toxin A is inhibited by lectins specific for Gal and GlcNAc, by immunoglobulin and nonimmunoglobulin components of human milk (Rolfe and Song 1995). It was reported that toxin A binds to the membranous sucrose-isomaltase glycoprotein on rabbit cells (Pothoulakis et al. 1996). This receptor is absent in many toxin A-sensitive cell lines. A human glycosphingolipid was reported to bind to toxin A (Teneberg et al. 1996). It is now generally accepted that a carbohydrate structure (containing at least Galβ1-4GlcN) is the essential element for binding of toxin A to its cell receptor; whether this carbohydrate is linked to proteins or lipids is unknown. So far there are no data on the receptors for the other glucosylating toxins.

2.6 Toxin Processing and Uptake

Subsequent to receptor binding, the toxins are endocytosed (Florin and Thelestam 1983; Henriques et al. 1987). It appears that the clostridial glucosylating toxins or their enzyme domains enter the cytosol from early endosomes (see below). Accordingly, it was shown that all drugs inhibiting acidification of endosomes, such as bafilomycin A1, block toxin entry and cytopathic effects of toxin B. Interestingly, the inhibitory effect of bafilomycin A1 can be bypassed by acidification of the culture medium (pH 5.2). This allows toxin translocation directly across the cell membrane into the cytosol (Barth et al. 2001; Qa'dan et al. 2000).

Recently, it was reported that toxin B is processed during its uptake (Pfeifer et al. 2003). After treatment of Vero cells with toxin B holotoxin, only an enzymatically active fragment but not the holotoxin B is detectable in the cytosol by immunoblot analysis, fluorescence microscopy, and mass spectrometry. Fluorescence microscopy data rather suggested that the translocation/binding domain remains in the endosomes (Pfeifer et al. 2003). At which step of the uptake process proteolytic cleavage of toxin B takes place is unclear. Proteolytic processing may occur at the cell surface, in endosomes, or even after translocation of the catalytic domain across the endosomal membrane by a cytosolic protease.

The central region of all clostridial glucosylating toxins carries a hydrophobic region (amino acid residues 1,000–1,100 in the toxins), which may be involved in translocation. Deduced from the uptake mechanism of other toxins, the acidic pH in endosomes is thought to trigger conformational changes in the hydrophobic stretch, thereby allowing membrane insertions and, eventually, translocation of the enzyme domain into the cytosol. This notion is supported by the finding that a decrease in pH increases hydrophobicity of toxin B, a conformational prerequisite for membrane insertion (Qa'dan et al. 2000).

Clostridium difficile toxin B and the binding and translocation domains (amino acids 547–2366) devoid of the catalytic part form pores in membranes of Chinese hamster ovary cells to pass small ions (Barth et al. 2001). As expected, the mere catalytic domain is not able to form pores. Furthermore, toxin B is also able to form pH-dependent channels in artificial bilayers. However, one should keep in mind that the role of the pore formation in translocation of the toxin is far from being clear. It is only speculation that the enzyme domain of the toxins is transported through this pore.

Clostridium sordellii lethal toxin, which consists of 2,364 amino acid residues with a mass of 270 kDa (Green et al. 1995), shares many biological properties with toxin A and B (Bette et al. 1991; Martinez and Wilkins 1988, 1992). The toxin is much less cytotoxic than toxin B and rather comparable to toxin A. However, it is 10 times more lethal than toxins A and B in mice after intraperitoneal injection. It shares glucosyltransferase activity with toxins A and B. One major difference between *Clostridium difficile* toxins A and B and lethal toxin is substrate specificity. It glucosylates Rac but not RhoA and, in addition, Ras subfamily GTPases. *Clostridium sordellii* hemorrhagic toxin (300 kDa) is very similar to toxin A, and it shares the enzyme properties of toxin A (Genth et al. 1996). Like toxin A it is able to elicit fluid response in ligated-loop assays (Martinez and Wilkins 1988, 1992). *Clostridium novyi* has been cloned and sequenced by the group of von Eichel-Streiber (Hofmann et al. 1995). It has a mass of about 250 kDa and consists of 2,178 amino acid residues. The α-toxin has biological properties very similar to those of toxins A and B (Ball et al. 1993; Bette et al. 1989, 1991; Oksche et al. 1992). The cytotoxic activity is similar to that of toxin B.

Recently, many forms of variant toxin B have been identified, which are functional chimeras of toxin B (reference strain VPI 10463) and lethal toxin, e.g., *Clostridium difficile* toxin B from strain 1470 and strain 8864. Their substrate specificities resemble that of lethal toxin (Chaves-Olarte et al. 1999; Rupnik et al. 1997, 1998; Soehn et al. 1998). *Clostridium difficile* strain C34 produces a toxin B variant modifying Rho, Rac, and Cdc42 as well as R-Ras, Ral, and Rap (Mehlig et al. 2001).

2.7
Glucosylation of Low-Molecular-Mass GTPases by Clostridial Glucosylating Toxins

All clostridial glucosylating toxins modify Rho A, B, and C at threonine-37 and Rac, Cdc42, or Ras at the corresponding position threonine-35 (Just et al. 1995b, 1995c, 1996) (Fig. 3). This amino acid residue, which is located in the switch-I region of the GTPases, is highly conserved among all small GTP-binding proteins. Rho is preferentially modified in its inactive, GDP-bound state, in which threonine-37 is directed toward the solvent. In the GTP-bound state, threonine-37 participates in Mg^{2+} and nucleotide binding (Ihara et al. 1998) and appears to be less accessible for glucosylating clostridial toxins (Just et al. 1995b).

The biochemical and functional consequences of toxin-catalyzed glucosylation of Rho/Ras proteins have been studied in great detail. Inhibition of effector coupling and subsequent blocking of signal transduction pathways is suggested to be the most important consequence of glycosylation (Herrmann et al. 1998; Sehr et al. 1998). Glucosylation of Rho proteins also blocks nucleotide exchange by GEFs and inhibits the intrinsic and GAP-stimulated GTPase activity (Sehr et al. 1998). It was further reported that glucosylated Rho is no longer able to interact with GDI and, therefore, is found at the plasma membrane, where it interacts with an 70-kDa protein of unknown nature (Genth et al. 1999). The glucosylated GTPase is entrapped at the membranes, however, without signaling because effector coupling is completely blocked. Every Rho-dependent pathway is inhibited. (Fig. 1B)

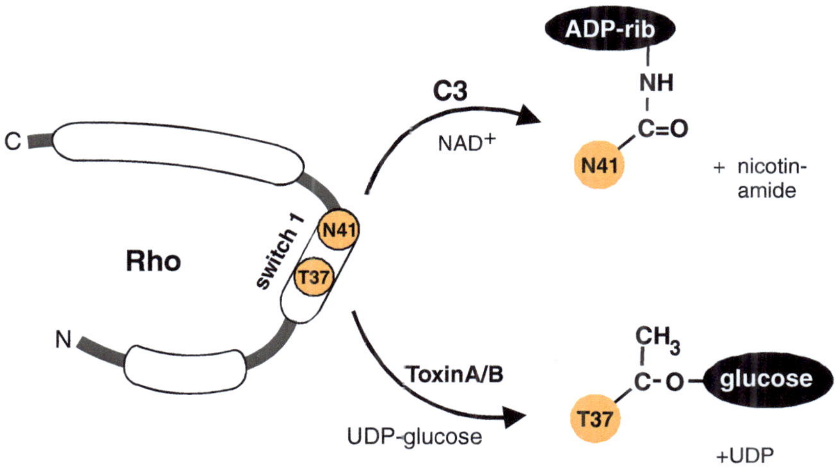

Fig. 3 Toxin-catalyzed Rho modifications. The acceptor amino acids for ADP-ribosylation and glucosylation reside in the effector region (switch 1). C3 forms a ternary complex with NAD^+ and RhoA and transfers the ADP-ribose moiety to the side chain of Asn-41; nicotinamide is released. ADP-ribose is *N*-glycosidically linked. Toxin B transfers a glucose moiety from UDP-glucose to the side chain of threonine-37, where the glucose is *O*-glycosidically linked; UDP is released. The proximity of Asn-41 and Thr-37 explains why one modification prevents the other

Functional consequences of *Clostridium sordellii* lethal toxin-mediated Ras glucosylation at threonine-35 are very similar to those of RhoA at threonine-37. Whereas nucleotide binding is not affected, the intrinsic GTPase activity is markedly decreased and the GAP-stimulated GTPase activity is completely blocked. Like glucosylated Rho proteins, the GEF-catalyzed GDP exchange to GTP is also reduced. Again, the most important consequence of Ras glucosylation appears to be inhibition of Ras coupling to downstream effectors. It was shown that the interaction of glucosylated Ras with the Ras-binding domain (RBD) of the Raf kinase is completely blocked. Kinetic studies exhibited a K_d value for the interaction of Ras-GTP with RafRBD of 15 nM, whereas glucosylation of Ras increased the K_d to >1 mM (Herrmann et al. 1998). These data were supported by crystal structure analysis of glucosylated and nonmodified Ras showing that glucosylation of the GTPases most likely blocks effector interaction. Moreover, the crystallographic data suggest that the glucose moiety hinders formation of the active GTP-bound conformation of the effector region, although mere binding of GTP is still possible (Vetter et al. 2000). Notably, threonine-35 of Ras, which is glucosylated by lethal toxin, is not directly involved in effector binding. As mentioned above, in the active form of Ras, the hydroxyl side chain of threonine-35 is directed into the core of the molecule and is involved in Mg^{2+} binding. Therefore, it was proposed that Ras glucosylation is solely possible in the GDP-bound form and Ras bound to GTP[S] is not substrate for lethal toxin (Herrmann et al. 1998). However, NMR analysis of soluble Ras bound to the GTP analog GppNHp suggested that the effector loop exists in two distinct conformational states that cycle rapidly (Geyer et al. 1996). Only in one of these conformational states is threonine-35 involved in Mg^{2+} binding; in the other state, however, it is probably accessible for glucosylation. After glucosylation the effector loop is stabilized in the inactive state. Altogether, NMR data suggest that glucosylation of GTP-bound Ras is possible (Geyer et al. 2003). NMR and crystal structure analysis support the view that glucose is bound in the α-anomeric form to the hydroxyl group of the threonine-35 side chain (Geyer et al. 2003; Vetter et al. 2000).

2.8 Effects of Glucosylating Toxins on Cells

Toxin-induced glucosylation of Rho/Ras proteins causes dramatic morphological changes in eukaryotic cells. A dramatic redistribution of the actin cytoskeleton takes place: Cells shrink and round up, an event that is initially accompanied by formation of neurite like retraction fibers. Finally, these fibers disappear and cells detach from the matrix (Fiorentini et al. 1989; Malorni et al. 1990; Ottlinger and Lin 1988; Thelestam and Chaves-Olarte 2000). Numerous cellular responses subsequent to inactivation of Rho and Ras proteins by glucosylation have been described; most of them are plausibly explained by inhibition of the multiple functions of the small GTPases modified. Reports include inhibition of phospholipase D activity (Schmidt et al. 1996), secretion (Prepens et al. 1996), phagocytosis (Caron and Hall 1998), calcium mobilization (Djouder et al. 2000), muscarinic receptor signaling to focal adhesion kinase (Linseman et al. 2000), as well as deregulation of neurotransmitter exocytosis (Doussau et al. 2000), apoptosis (Subauste et al. 2000), chemoattractant receptor signaling (Servant et al. 2000), and neuronal axon formation (Bradke and Dotto 1999).

2.9 Pathophysiological Role of *Clostridium difficile* Toxins

Although our knowledge about the structure and functions of glucosylating toxins from *Clostridium difficile* has increased enormously, the precise pathogenetic pathways that finally result in toxin-induced diarrhea and pseudomembranous colitis are still not clear. Toxin-induced fluid response and diarrhea may be explained by tissue damage and inhibition of the barrier function of the enterocytes. Many studies have reported that the toxins have major effects on tight junctions. They decrease transepithelial resistance, increase paracellular bacterial migration, and change the morphological features of tight junctions and associated proteins (Feltis et al. 2000; Gerhard et al. 1998; Hecht et al. 1988, 1992; Liu et al. 2003; Moore et al. 1990; Nusrat et al., 2001; Riegler et al. 1995). For example, F-actin restructuring induced by the toxins is accompanied by dissociation of occludin, ZO-1, and ZO-2 from lateral tight junction without affecting adherence junctions. These data are in line with the view that Rho GTPases play a pivotal role in tight junction regulation (Jou et al. 1998; Nusrat et al. 1995).

In addition to altered barrier function of enterocytes, toxins A and B induce a pathological feature in the gut that may be summarized as a major inflammatory response. The toxins induce massive neutrophil infiltration and the production and release of various inflammatory mediators, including prostaglandins, and leukotrienes (Mahida et al. 1996; Pothoulakis et al. 1988), Il-8 (He et al. 2002; Jefferson et al. 1999; Mahida et al. 1996; Savidge et al. 2003; Warny et al. 2000), and TNF-α (Ishida et al. 2004). Toxin A reportedly activates nuclear factor NF-κB. In rat intestine macrophages are activated to release MIP-2 (Pothoulakis and Lamont 2001). Furthermore, the toxins may activate intestinal nerves to release neuropeptides substance P and calcitonin gene-related peptide (CGRP) (Pothoulakis et al. 1994), which have proinflammatory properties. A specific role is assigned to mast cells, which appear to be degranulated early after toxin A exposure (Wershil et al. 1998). Although Rho GTPases are crucially involved in regulation of immune actions and transcriptional activation of immune cells, it was suggested that some of these responses are independent of Rho GTPases. It was reported that toxin A causes ZO-1 translocation and increases paracellular flux via protein kinase C signal pathways in a process that occurs earlier than glucosylation of Rho proteins by the toxin (Chen et al. 2002). In addition, mitochondrial damage and p38 mitogen activation were reported to be independent of Rho.

3
C3-Like ADP-Ribosyltransferases

Clostridium botulinum C3 ADP-ribosyltransferase was discovered in the 1980s during the course of screening for higher producer strains of the actin-ADP-ribosylating C2 toxin (Aktories et al. 1987). C3 is a monomeric 24-kDa single-chain peptide that has an enzyme domain but apparently no binding and translocation domain. Thus the transferase is designated as an exoenzyme but not as a toxin. Later, related C3 exoenzymes were found to be produced by *Clostridium limosum* (C3lim), *Bacillus cereus* (C3cer), and *Staphylococcus aureus* (C3stau1, 2, 3, also called EDINs), which share 30% to 60% identity at the amino acid level and thus have been joined together as the "family of C3-like exoenzymes" (Inoue et al. 1991; Just et al. 1992, 1995a; Rubin et al. 1988; Sugai et al. 1990; Wilde et al. 2001b; Yamaguchi et al. 2001). C3 exoenzymes are secreted by the producing microbes, and thus all of them contain a signal sequence. They catalyze the transfer of an ADP-ribose moiety from the cosubstrate NAD^+ to the RhoA/B/C-GTPases and covalently link it *N*-glycosidically (Fig. 3).

3.1 Structure–Function Analysis of C3 Exoenzymes

The crystal structure of C3 exoenzyme from *Clostridium botulinum* (C3bot1) (Han et al. 2001; Ménétrey et al. 2002) and from the related C3stau (Evans et al. 2003) has been solved. C3 exoenzymes share the typical folding of ADP-ribosyltransferases. The core of the enzymes consists of a five-stranded mixed β-sheet, which is positioned against a three-stranded antiparallel β-sheet (Fig. 4). Four α-helices flank the three-stranded β-sheet. An additional α-helix flanks the five-stranded β-sheet. The catalytic pocket, including the NAD-binding site, is formed by the β-sheet core and one α-helix ($\alpha 3$).

C3 exoenzymes share a number of conserved amino acids in the active site between themselves and also with other ADP-ribosyltransferases, especially those from the group of actin-ADP-ribosylating toxins. Glu-214 in C3bot1 (numbering with signal sequence) is the *catalytic* glutamic acid residue, which is conserved not only in all C3 exoenzymes but also in all ADP-ribosyltransferases studied so far. Changes of this catalytic Glu to Asp or Gln in C3lim or to any amino acid in C3bot strongly reduces ADP-ribosyltransferase activity and leads to a decrease in the affinity of the cosubstrate NAD^+ (Böhmer et al. 1996; Saito et al. 1995).

All C3-like exoenzymes ADP-ribosylate Rho GTPases (e.g., RhoA, B, and C) at Asn-41 (Sekine et al. 1989). The side chain of Asn-41 is solvent exposed and thus accessible, as can be deduced from the RhoA crystal structure (Ihara et al. 1998; Wei et al. 1997). Because the accessibility is not changed by nucleotide binding, Rho-GTP as well as Rho-GDP are substrates for C3 (Inoue et al. 1991; Just et al. 1992, 1995a; Sugai et al. 1990; Wilde et al. 2001b; Yamaguchi et al. 2001). It was suggested that a motif in C3bot termed "ARTT" (ADP-

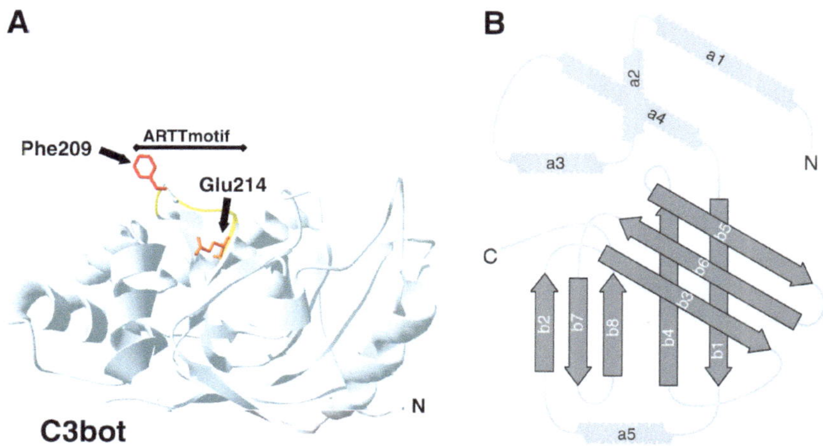

Fig. 4A, B. Structure of C3bot. **A** The ADP-ribosylation toxin-turn-turn (*ARTT*) motif is shown with the "catalytic glutamate" (*Glu 214 for C3bot*) and phenylalanine at position 209, which might be involved in protein substrate binding. Data by Swiss-Pdb Viewer 3.7 (Database code 1G24). **B** Scheme of the folding of C3bot (see text)

ribosylating-toxin-Turn-Turn motif) is crucial for the recognition of the RhoA (Han et al. 2001) (Fig. 4). The ARTT motif consists of two stretches covering residues Ser-207 to Ala-210 (Turn 1) and Gly-211 to Glu-214 (Turn 2) of C3bot. Both "turns" are located close to the N1 and the C1′ atoms of NAD, which form the *N*-glycosidic bond. The second "turn" of the ARTT-motif, which contains the solvent-exposed Gln-212, is suggested to interact with the carbonyl and amide groups of Asn-41 of RhoA. The first ARTT motif with residue Phe-209 of C3bot (Phe in C3lim) is believed to function in the recognition of RhoA by interacting with the hydrophobic, solvent-exposed patches around Rho Asn41.

3.2
Functional Consequences of ADP-Ribosylation of Rho Proteins by C3 ADP-Ribosyltransferases

C3-like exoenzymes ADP-ribosylate RhoA, B, and C at Asn-41 (Aktories et al. 1989; Braun et al. 1989; Chardin et al. 1989; Sekine et al. 1989). Although most other members of the Rho GTPase family contain an asparagine at the same position, they are poor or no *in vitro* substrates (e.g., Rac) (Just et al. 1992). The acceptor residue Asn-41 is located in the switch-1 region (residues 28–41) of the GTPase (Ihara et al. 1998; Wei et al. 1997). This region undergoes major conformational changes depending on the nucleotide binding state. Asn-41 is in the vicinity of Thr-37, which is glucosylated by *Clostridium difficile* toxins A and B, and glucosylation inhibits C3-catalyzed ADP-ribosylation. Conversely, ADP-ribosylation at Asn-41 blocks subsequent glucosylation at Thr-37 (Fig. 1).

ADP-ribosylation of RhoA renders the GTPase biologically inactive. C3 exoenzymes cause a dramatic redistribution of the actin cytoskeleton characteristic for the inactivation of RhoA (Chardin et al. 1989; Paterson et al. 1990; Wiegers et al. 1991). ADP-ribosylation has only minimal effects on nucleotide binding and GTP-hydrolyzing activity of Rho (Paterson et al. 1990). However, ADP-ribosylated RhoA is still able to interact with various effector proteins such as protein kinase N, Rho kinase, and phospholipase D (Genth et al. 2003a, 2003b; Sehr et al. 1998). This property is in clear contrast to Rho glucosylated at Thr-37 by *Clostridium difficile* toxin B. Because binding of Rho to effector kinases is thought to be sufficient for activation of the kinase activity, blockade of the Rho-effector interaction is not the mechanism underlying inactivation of Rho signaling by ADP-ribosylation. ADP-ribosylation of Rho may even increase the affinity toward its effector (e.g., phosphatidylinositol-4-phosphate-5-kinase) (Ren et al. 1996), possibly sequestering the effector. ADP-ribosylation also inhibits the activation of RhoA by GEF (e.g., Lbc) (Sehr et al. 1998). Moreover, ADP-ribosylated RhoA accumulates in the cytoplasm and is entrapped in the GDI complex, and the release from GDI appears to be blocked (Fujihara et al. 1997; Genth et al. 2003a). Because Rho requires a translocation from the cytoplasm to the membranes to become activated, translocation and activation of Rho will not occur after ADP-ribosylation. Despite the property of ADP-ribosylated RhoA to bind to effector proteins in a recombinant system, in the cellular context ADP-ribosylation stabilizes the inactive state of Rho, thereby preventing activation (Fujihara et al. 1997; Genth et al. 2003a).

In addition to the direct interference with GTPase and cytosol membrane cycling, ADP-ribosylation seems to make RhoA sensitive to proteolytic degradation. It is conceivable that through decrease in cellular Rho concentration the cross talk between Rho, Rac, and Cdc42 is disturbed, resulting in additional cellular effects (Barth et al. 1999; Malcolm et al. 1996; Meacci et al. 1999).

3.3
Additional Targets of C3 Exoenzymes

RhoA, B, and C are the canonical intracellular target proteins of the C3-like exoenzymes. However, the C3 isoforms from *S. aureus* (designated C3stau) are able to additionally ADP-ribosylate Rnd/RhoE proteins (Wilde et al. 2001b). The acceptor amino acid is Asn-44 in RhoE/Rnd3, which is equivalent to Asn-41 in RhoA. Rnd GTPases belong also to the Rho GTPase family, but they exhibit a special property: They do not possess GTP-hydrolyzing activity, and they are therefore permanently active. Rnd3 GTPase is a functional antagonist of RhoA leading to reorganization of the actin cytoskeleton (Wennerberg and Der 2004).

The C3 isoforms of clostridial and bacillus origin interact with an additional target protein, however, without modification. This target protein is the Ral GTPase, which belongs to the Ras GTPase family. Ral is in fact not ADP-ribosylated, but it is bound with high affinity to C3. This strong binding inhibits Ral signaling activity, for example, to regulate PLD, and conversely, transferase activity of C3 is blocked (Wilde et al. 2002). Thus it is conceivable that C3 exhibits effects other than inactivation of RhoA, especially in those cells expressing a lot of Ral, such as neuronal cells.

3.4
Cellular Effects of C3-Like ADP-Ribosyltransferases

Many studies on Rho functions were performed with C3 transferases. Only some examples of studies are reviewed here, which show the successful application of C3 exoenzymes as pharmacological and cell biological tools.

In Vero cells, C3bot (5 μg/ml, 12–24 h) induces morphological changes characterized by rounding up of the cells with concomitant destruction of stress fibers (Chardin et al. 1989). Similar findings were obtained with many cell types and with different types of C3 exoenzymes (Barth et al. 1998; Paterson et al. 1990; Ridley and Hall 1992, 1994; Wilde et al. 2001b). A common observation is that after C3 treatment cell-cell contact remains via small extensions, whereas treatment with the actin-ADP-ribosylating C2 toxin causes complete loss of cell contacts (Wiegers et al. 1991). Whereas cortical actin appears to be more resistant toward C3 treatment, loss of stress fibers caused by inactivation of RhoA is typical. Usually cells are still viable after C3 treatment. Exchange of the medium to remove C3 reverses cell rounding after a few hours or days, probably because of neosynthesis of Rho (Barth et al. 1999). After microinjection the effects are more rapidly observed and occur within 10–15 min (Paterson et al. 1990). The pioneering studies on Rho GTPases performed in the laboratory of Alan Hall must be mentioned here. In many of these studies C3 was crucial for identification of the role of Rho proteins in organization of the actin cytoskeleton on extracellular stimuli (Hall 1994; Mackay et al. 1997; Ridley and Hall 1992). Accordingly, processes that are suggested to be mediated by Rac or Cdc42 are not affected by C3, underlining the substrate specificity of the transferases (Kozma et al. 1995; Nobes and Hall 1995; Ridley et al. 1992). C3 was successfully applied in studies on the role of Rho GTPases in neurite outgrowth (Jalink et al. 1994; Mackay et al. 1995; Tigyi et al. 1996; Wahl et al. 2000) and in studies on the role of Rho in endocytosis (Lamaze et al. 1996) and phagocytosis (Caron and Hall 1998; Lamaze et al. 1996; Schmalzing et al. 1995; Vögler et al. 1999). Using C3, the involvement of Rho proteins in the regulation of the phosphatidylinositol-4-phosphate-5 kinase (PI-4P-5-kinase) and phospholipase D was studied (Balboa and Insel 1995; Kuribara et al. 1995; Meacci et al. 1999; Schmidt et al. 1996, 1999; Weernick et al. 2000). C3 was applied to investigate the role of Rho in signaling to the nucleus and in regulation of gene transcription (Alberts et al. 1998; Hill et al. 1995). Moreover, C3bot was successfully employed in delineation of the role of Rho in signal transduction from heptahelical receptors to the nucleus via heterotrimeric G proteins (Fromm et al. 1997; Mao et al. 1998a, 1998b; Sah et al. 1996).

3.5
C3 Exoenzymes Are Pharmacological Tools

C3 is frequently applied as a pharmacological tool to inactivate RhoA, B, and C, in order to study the functional roles of Rho in signaling processes. Because C3 transferases only consist of the catalytic domain without a cell binding and transport domain, cells are poorly accessible for C3. Studies with intact cells require high concentrations (e.g., 5–50 μg/ml) and long incubation times (up to 24–48 h) (Amano et al. 1996; Morii and Narumiya 1995; Verschueren et al. 1997; Wiegers et al. 1991). Often the toxin is microinjected (Chong et al. 1994; Olson et al. 1998; Paterson et al. 1990; Ridley and Hall 1992, 1994; Watanabe et al. 1997). Other methods to introduce the toxin into cultured cells include permeabilization of cells by digitonin (Mackay et al. 1997), streptolysin O (Fensome et al. 1998), electropermeabilization (Koch et al. 1994; Stasia et al. 1991) or by scrape loading (Barry et al. 1997). The C3 gene was introduced into eukaryotic cells by transient and stable transfection with plasmids or by viral infection, and even transgenic mice have been developed with thymocyte- or lens-specific expression of the C3 exoenzyme (Caron and Hall 1998; Fujisawa et al. 1998; Genot et al. 1996; Henning et al. 1997; Hill et al. 1995; Maddala et al. 2004; Meacci et al. 1999).

To improve cell accessibility, chimeras were constructed, consisting of C3 ADP-ribosyltransferases and the cell binding/translocation domain of "complete" AB toxins. In one approach, C3bot was fused to the binding and translocation subunit of diphtheria toxin (Aullo et al. 1993). Recently, the *Clostridium botulinum* C2 toxin was used to construct a chimeric fusion toxin (Barth et al. 1998, 2002; Meyer et al. 2000). The chimeric C3 toxins allow application at low concentration because a specific uptake process is used. Furthermore, the incubation time can reasonably be reduced to hours compared to days when wild-type C3 is applied.

3.6 The Role of C3 as a Virulence Factor

A pathophysiological action of C3-like ADP-ribosyltransferases on the immune system of the target organism is obvious. This view is supported by many reports, showing that C3 affects immune cell functions (Lang et al. 1992; Laudanna et al. 1996, 1997; Moss et al. 1997; Nemoto et al. 1996; Stam et al. 1998). Rho GTPases regulate processes important for host immune responses (Henning and Cantrell 1998; Laudanna et al. 1996; Reif and Cantrell 1998; Wojciak-Stothard et al. 1998) and participate in the barrier functions of epithelial cells (Nusrat et al. 1995; Vouret-Craviari et al. 1998) and in wound healing (Santos et al. 1997). How C3-like transferases reach their intracellular target proteins is still an open question. It has been suggested for C3stau-producing *S. aureus* that they are able to invade cells to survive intracellularly with a technique that allows them to escape the phagosomes (Wilde et al. 2001a). Release of the transferase would then occur at the place where Rho GTPases are located, without further need for membrane crossing. On the other hand, it was proposed that membrane-damaging bacterial toxins facilitate the cellular entry of other bacterial enzymes and effectors (Madden et al. 2001). If this is also true for *Clostridia*, it is feasible that C3 enters cells through the help of hemolysins or other pore-forming toxins, which are produced by many *Clostridia*, including those producing C3.

3.7 Nonenzymatic Effects of C3

The effect of C3bot on neurite outgrowth has been explained by its Rho inactivating property. However, it was reported recently that the effect of C3bot to induce axonal growth was independent of its inherent transferase activity (Ahnert-Hilger et al. 2004). Enzymatically deficient C3bot showed neurotrophic effects. This property is unique for C3bot and is not shared by the other members of the C3-like family from *C. limosum* and *S. aureus* (Ahnert-Hilger et al. 2004).

4
Conclusions

Discovery of and research on clostridial Rho-inactivating protein toxins have had a significant impact on the understanding of the biology of Rho GTPases. In this respect, all the studies mentioned above reflect the enormous advances that have been made during recent years in the novel field of "cellular microbiology." Nevertheless, many important questions remain: For example, the structure-function relationships of the clostridial glucosylating toxins are far from being understood. So far we do not have any crystal structure of the toxins, and, most likely, we will have to wait even longer for the structure of the toxins together with their substrates. Further progress in the field will certainly also boost the understanding of the role of GTPases in diseases induced by the toxin-producing pathogens.

References

Ahnert-Hilger G, Holtje M, Grosse G, Pickert G, Mucke C, Nixdorf-Bergweiler B, Boquet P, Hofmann F, Just I (2004) Differential effects of Rho GTPases on axonal and dendritic development in hippocampal neurones. J Neurochem 90:9–18

Aktories K, Braun U, Rösener S, Just I, Hall A (1989) The rho gene product expressed in *E. coli* is a substrate of botulinum ADP-ribosyltransferase C3. Biochem Biophys Res Commun 158:209–213

Aktories K, Weller U, Chhatwal G S (1987) *Clostridium botulinum* type C produces a novel ADP-ribosyltransferase distinct from botulinum C2 toxin. FEBS Lett 212:109–113

Alberts A S, Geneste O, Treisman R (1998) Activation of SRF-regulated chromosomal templates by Rho-family GTPases requires a signal that also induces H4 hyperacetylation. Cell 92:475–487

Amano M, Mukai H, Ono Y, Chihara K, Matsui T, Hamajima Y, Okawa K, Iwamatsu A, Kaibuchi K (1996) Identification of a putative target for Rho as the serine-threonine kinase protein kinase N. Science 271:648–650

Aullo P, Giry M, Olsnes S, Popoff M R, Kocks C, Boquet P (1993) A chimeric toxin to study the role of the 21 kDa GTP binding protein rho in the control of actin microfilament assembly. EMBO J 12:921–931

Balboa M A, Insel P A (1995) Nuclear phospholipase D in Madin-Darby canine kidney cells—Guanosine 5′-O-(thiotriphosphate)-stimulated activation is mediated by RhoA and is downstream of protein kinase C. J Biol Chem 270:29843–29847

Ball D W, Van Tassell R L, Denton Roberts M, Hahn P E, Lyerly D M, Wilkins T D (1993) Purification and characterization of α-toxin produced by *Clostridium novyi* type A. Infect Immun 61:2912–2918
Barbieri J T, Riese M J, Aktories K (2002) Bacterial toxins that modify the actin cytoskeleton. Annu Rev Cell Dev Biol 18:315–344
Barroso L A, Moncrief J S, Lyerly D M, Wilkins T D (1994) Mutagenesis of the *Clostridium difficile* toxin B gene and effect on cytotoxic activity. Microb Pathog 16:297–303
Barroso L A, Wang S-Z, Phelps C J, Johnson J L, Wilkins T D (1990) Nucleotide sequence of *Clostridium difficile* toxin B gene. Nucl Acids Res 18:4004-
Barry S T, Flinn H M, Humphries M J, Critchley D R, Ridley A J (1997) Requirement for Rho in integrin signalling. Cell Adhes Commun 4:387–398
Barth H, Hofmann F, Olenik C, Just I, Aktories K (1998) The N-terminal part of the enzyme component (C2I) of the binary *Clostridium botulinum* C2 toxin interacts with the binding component C2II and functions as a carrier system for a Rho ADP-ribosylating C3-like fusion toxin. Infect Immun 66:1364–1369
Barth H, Olenik C, Sehr P, Schmidt G, Aktories K, Meyer D K (1999) Neosynthesis and activation of Rho by *Escherichia coli* cytotoxic necrotizing factor (CNF1) reverse cytopathic effects of ADP-ribosylated Rho. J Biol Chem 274:27407–27414
Barth H, Pfeifer G, Hofmann F, Maier E, Benz R, Aktories K (2001) Low pH-induced formation of ion channels by *Clostridium difficile* toxin B in target cells. J Biol Chem 276:10670–10676
Barth H, Roebling R, Fritz M, Aktories K (2002) The binary *Clostridium botulinum* C2 toxin as a protein delivery system. J Biol Chem 277:5074–5081
Bartlett J G (2002) Clinical Practice: Antibiotic-associated Diarrhea. N Engl J Med 346:334–339
Bartlett J G, Moon N, Chang T W, Taylor N, Onderdonk A B (1978) Role of *Clostridium difficile* in antibiotic-associated pseudomembranous colitits. Gastroenterology 75:778–782
Bette P, Frevert J, Mauler F, Suttorp N, Habermann E (1989) Pharmacological and biochemical studies of cytotoxicity of *Clostridium novyi* type A α-toxin. Infect Immun 57:2507–2513
Bette P, Oksche A, Mauler F, Von Eichel-Streiber C, Popoff M R, Habermann E (1991) A comparative biochemical, pharmacological and immunological study of *Clostridium novyi* α-toxin, *C. difficile* toxin B and *C. sordellii* lethal toxin. Toxicon 29:877–887
Bishop A L, Hall A (2000) Rho GTPases and their effector proteins. Biochem J 348:241–255
Böhmer J, Jung M, Sehr P, Fritz G, Popoff M, Just I, Aktories K (1996) Active site mutation of the C3-like ADP-ribosyltransferase from *Clostridium limosum*—Analysis of glutamic acid 174. Biochemistry 35:282–289
Bradke F, Dotto G P (1999) The role of local actin instability in axon formation. Science 283:1931–1934
Braun U, Habermann B, Just I, Aktories K, Vandekerckhove J (1989) Purification of the 22 kDa protein substrate of botulinum ADP-ribosyltransferase C3 from porcine brain cytosol and its characterization as a GTP-binding protein highly homologous to the rho gene product. FEBS Lett 243:70–76
Breton C, Imberty A (1999) Structure/function studies of glycosyltransferases. Curr Opin Struct Biol 9:563–571
Busch C, Aktories K (2000) Microbial toxins and the glucosylation of Rho family GTPases. Curr Opin Struct Biol 10:528–535

Busch C, Hofmann F, Gerhard R, Aktories K (2000a) Involvement of a conserved tryptophan residue in the UDP-glucose binding of large clostridial cytotoxin glycosyltransferases. J Biol Chem 275:13228–13234

Busch C, Hofmann F, Selzer J, Munro J, Jeckel D, Aktories K (1998) A common motif of eukaryotic glycosyltransferases is essential for the enzyme activity of large clostridial cytotoxins. J Biol Chem 273:19566–19572

Busch C, Schömig K, Hofmann F, Aktories K (2000b) Characterization of the catalytic domain of *Clostridium novyi* α-toxin. Infect Immun 68:6378–6383

Caron E, Hall A (1998) Identification of two distinct mechanisms of phagocytosis controlled by different Rho GTPases. Science 282:1717–1721

Chardin P, Boquet P, Madaule P, Popoff M R, Rubin E J, Gill D M (1989) The mammalian G protein rho C is ADP-ribosylated by *Clostridium botulinum* exoenzyme C3 and affects actin microfilament in Vero cells. EMBO J 8:1087–1092

Chaves-Olarte E, Löw P, Freer E, Norlin T, Weidmann M, Von Eichel-Streiber C, Thelestam M (1999) A novel cytotoxin from *Clostridium difficile* serogroup F is a functional hybrid between two other large clostridial cytotoxins. J Biol Chem 274:11046–11052

Chen M L, Pothoulakis C, LaMont J T (2002) Protein kinase C signaling regulates ZO-1 translocation and increased paracellular flux of T84 coloncytes exposed to *Clostridium difficile* toxin A. J Biol Chem 277:4247–4254

Chong L D, Traynor-Kaplan A, Bokoch G M, Schwartz M A (1994) The small GTP-binding protein Rho regulates a phosphatidylinositol 4-phosphate 5-kinase in mammalian cells. Cell 79:507–513

Ciesla W P, Jr., Bobak D A (1998) Clostridium difficile toxins A and B are cation-dependent UDP-glucose hydrolases with differing catalytic activities. J Biol Chem 273:16021–16026

Djouder N, Prepens U, Aktories K, Cavalié A (2000) Inhibition of calcium release-activated calcium current by Rac/Cdc42-inactivating clostridial cytotoxins in RBL cells. J Biol Chem 275:18732–18738

Doussau F, Gasman S, Humeau Y, Vitiello F, Popoff M, Boquet P, Bader M-F, Poulain B (2000) A Rho-related GTPase is involved in Ca^{2+}-dependent neurotransmitter exocytosis. J Biol Chem 275:7764–7779

Dove C H, Wang S Z, Price S B, Phelps C J, Lyerly D M, Wilkins T D, Johnson J L (1990) Molecular characterization of the *Clostridium difficile* toxin A gene. Infect Immun 58:480–488

Etienne-Manneville S, Hall A (2002) Rho GTPases in cell biology. Nature 420:629–635

Evans H R, Sutton J M, Holloway D E, Ayriss J, Shone C C, Acharya K R (2003) The crystal structure of C3stau2 from *Staphylococcus aureus* and its complex with NAD. J Biol Chem 278:45924–45930

Feltis B A, Wiesner S M, Kim A S, Erlandsen S L, Lyerly D L, Wilkins T D, Wells C L (2000) Clostridium difficile toxins A and B can alter epithelial permeability and promote bacterial paracellular migration through HT-29 enterocytes. Shock 14:629–634

Fensome A, Whatmore J, Morgan C, Jones D, Cockcroft S (1998) ADP-ribosylation factor and Rho proteins mediate fMLP-dependent activation of phospholipase D in human neutrophils. J Biol Chem 273:13157–13164

Fiorentini C, Arancia G, Paradisi S, Donelli G, Giuliano M, Piemonto F, Mastrantonio P (1989) Effects of *Clostridium difficile* toxins A and B on cytoskeleton organization in HEp-2 cells: a comparative morphological study. Toxicon 27:1209–1218

Flatau G, Lemichez E, Gauthier M, Chardin P, Paris S, Fiorentini C, Boquet P (1997) Toxin-induced activation of the G protein p21 Rho by deamidation of glutamine. Nature 387:729–733

Florin I, Thelestam M (1983) Internalization of *Clostridium difficile* cytotoxin into cultured human lung fibroblasts. Biochim Biophys Acta 763:383–392

Frey S M, Wilkins T D (1992) Localization of two epitopes recognized by monoclonal antibody PCG-4 on *Clostridium difficile* toxin A. Infect Immun 60:2488–2492

Frisch C, Gerhard R, Aktories K, Hofmann F, Just I (2003) The complete receptor-binding domain of *Clostridium difficile* toxin A is required for endocytosis. Biochem Biophys Res Commun 300:706–711

Fromm C, Coso O A, Montaner S, Xu N, Gutkind J S (1997) The small GTP-binding protein Rho links G protein-coupled receptors and $G\alpha_{12}$ to the serum response element and to cellular transformation. Proc Natl Acad Sci USA 94:10098–10103

Fu Y, Galán J E (1999) A *Salmonella* protein antagonizes Rac-1 and Cdc42 to mediate host-cell recovery after bacterial invasion. Nature 401:293–297

Fujihara H, Walker L A, Gong M C, Lemichez E, Boquet P, Somlyo A V, Somlyo A P (1997) Inhibition of RhoA translocation and calcium sensitization by in vivo ADP-ribosylation with the chimeric toxin DC3B. Mol Biol Cell 8:2437–2447

Fujisawa K, Madaule P, Ishizaki T, Watanabe G, Bito H, Saito Y, Hall A, Narumiya S (1998) Different regions of Rho determine Rho-selective binding of different classes of Rho target molecules. J Biol Chem 273:18943–18949

Genot E, Cleverley S, Henning S, Cantrell D (1996) Multiple p21ras effector pathways regulate nuclear factor of activated T cells. EMBO J 15:3923–3933

Genth H, Aktories K, Just I (1999) Monoglucosylation of RhoA at threonine-37 blocks cytosol-membrane cycling. J Biol Chem 274:29050–29056

Genth H, Gerhard R, Maeda A, Amano M, Kaibuchi K, Aktories K, Just I (2003a) Entrapment of Rho ADP-ribosylated by *Clostridium botulinum* C3 exoenzyme in the Rho-GDI-1 complex. J Biol Chem (in press)

Genth H, Hofmann F, Selzer J, Rex G, Aktories K, Just I (1996) Difference in protein substrate specificity between hemorrhagic toxin and lethal toxin from *Clostridium sordellii*. Biochem Biophys Res Commun 229:370–374

Genth H, Schmidt M, Gerhard R, Aktories K, Just I (2003b) Activation of phospholipase D1 by ADP-ribosylated RhoA. Biochem Biophys Res Commun 302:127–132

Gerhard R, Schmidt G, Hofmann F, Aktories K (1998) Activation of Rho GTPases by *Escherichia coli* cytotoxic necrotizing factor 1 increases intestinal permeability in Caco-2 cells. Infect Immun 66:5125–5131

Geyer M, Schweins T, Herrmann C, Prisner T, Wittinghofer A, Kalbitzer H R (1996) Conformational transitions in $p21^{ras}$ and in its complexes with the effector protein Raf-RBD and the GTPase activating protein GAP. Biochemistry 35:10308–10320

Geyer M, Wilde C, Selzer J, Aktories K, Kalbitzer H R (2003) Glucosylation of Ras by *Clostridium sordellii* lethal toxin: Consequences for the effector loop conformations observed by NMR spectroscopy. Biochemistry 42:11951–11959

Goehring U-M, Schmidt G, Pederson K J, Aktories K, Barbieri J T (1999) The N-terminal domain of *Pseudomonas aeruginosa* exoenzyme S is a GTPase-activating protein for Rho GTPases J Biol Chem 274:36369–36372

Green G A, Schué V, Monteil H (1995) Cloning and characterization of the cytotoxin L-encoding gene of *Clostridium sordellii*: Homology with *Clostridium difficile* cytotoxin B. Gene 161:57–61

Hall A (1994) Small GTP-binding proteins and the regulation of the actin cytoskeleton. Annu Rev Cell Biol 10:31–54

Han S, Arvai A S, Clancy S B, Tainer J A (2001) Crystal structure and novel recognition motif of Rho ADP-ribosylating C3 exoenzyme from *Clostridium botulinum*: Structural insights for recognition specificity and catalysis. J Mol Biol 305:95–107

He D, Sougioultzis S, Hagen S, Liu J, Keates S, Keates A C, Pothoulakis C, LaMont J T (2002) *Clostridium difficile* toxin triggers human colonocyte IL-8 release via mitochondrial oxygen radical generation. Gastroenterology 122:1048–1057

Hecht G, Koutsouris A, Pothoulakis C, LaMont J T, Madara J L (1992) *Clostridium difficile* toxin B disrupts the barrier function of T_{84} monolayers. Gastroenterology 102:416–423

Hecht G, Pothoulakis C, LaMont J T, Madara J L (1988) *Clostridium difficile* toxin A perturbs cytoskeletal structure and tight junction permeability of cultured human intestinal epithelial monolayers. J Clin Invest 82:1516–1524

Henning S W, Cantrell D A (1998) GTPases in antigen receptor signalling. Curr Opin Immunol 10:322–329

Henning S W, Galandrini R, Hall A, Cantrell D A (1997) The GTPase Rho has a critical regulatory role in thymus development. EMBO J 16:2397–2407

Henriques B, Florin I, Thelestam M (1987) Cellular internalisation of *Clostridium difficile* toxin A. Microb Pathogen 2:455–463

Herrmann C, Ahmadian M R, Hofmann F, Just I (1998) Functional consequences of monoglucosylation of H-Ras at effector domain amino acid threonine-35. J Biol Chem 273:16134–16139

Hill C S, Wynne J, Treisman R (1995) The Rho family GTPases RhoA, Rac1, and CDC42Hs regulate transcriptional activation by SRF. Cell 81:1159–1170

Hoffmann C, Pop M, Leemhuis J, Schirmer J, Aktories K, Schmidt G (2004) The *Yersinia pseudotuberculosis* cytotoxic necrotizing factor (CNFY) selectively activates RhoA. J Biol Chem 279:

Hofmann F, Busch C, Aktories K (1998) Chimeric clostridial cytotoxins: identification of the N-terminal region involved in protein substrate recognition. Infect Immun 66:1076–1081

Hofmann F, Busch C, Prepens U, Just I, Aktories K (1997) Localization of the glucosyltransferase activity of *Clostridium difficile* toxin B to the N-terminal part of the holotoxin. J Biol Chem 272:11074–11078

Hofmann F, Herrmann A, Habermann E, Von Eichel-Streiber C (1995) Sequencing and analysis of the gene encoding the α-toxin of *Clostridium novyi* proves its homology to toxins A and B of *Clostridium difficile*. Mol Gen Genet 247:670–679

Hofmann F, Rex G, Aktories K, Just I (1996) The Ras-related protein Ral is monoglucosylated by *Clostridium sordellii* lethal toxin. Biochem Biophys Res Commun 227:77–81

Ihara K, Muraguchi S, Kato M, Shimizu T, Shirakawa M, Kuroda S, Kaibuchi K, Hakoshima T (1998) Crystal structure of human RhoA in a dominantly active form complexed with a GTP analogue. J Biol Chem 273:9656–9666

Inoue S, Sugai M, Murooka Y, Paik S-Y, Hong Y-M, Ohgai H, Suginaka H (1991) Molecular cloning and sequencing of the epidermal cell differentiation inhibitor gene from *Staphylococcus aureus*. Biochem Biophys Res Commun 174:459–464

Ishida Y, Maegawa T, Kondo T, Kimura A, Iwakura Y, Nakamura S, Mukaida N (2004) Essential involvement of IFN-γ in *Clostridium difficile* toxin A-induced enteritis. J Immunol 172:3018–3025

Jalink K, Van Corven E J, Hengeveld T, Morii N, Narumiya S, Moolenaar W H (1994) Inhibition of lysophosphatidate- and thrombin-induced neurite retraction and neuronal cell rounding by ADP ribosylation of the small GTP-binding protein Rho. J Cell Biol 126:801–810

Jefferson K K, Smith M F Jr, Bobak D A (1999) Roles of intracellular calcium and NF-κB in the *Clostridium difficile* toxin A-induced up-regulation and secretion of IL-8 from human monocytes. J Immunol 163:5183–5191

Jou T-S, Schneeberger E E, Nelson W J (1998) Structural and functional regulation of tight junctions by RhoA and Rac1 small GTPases. J Cell Biol 142:101–115

Juris S J, Shao F, Dixon J E (2002) *Yersinia* effectors target mammalian signalling pathways. Cell Microbiol 4:201–211

Just I, Mohr C, Schallehn G, Menard L, Didsbury J R, Vandekerckhove J, van Damme J, Aktories K (1992) Purification and characterization of an ADP-ribosyltransferase produced by *Clostridium limosum*. J Biol Chem 267:10274–10280

Just I, Selzer J, Hofmann F, Green G A, Aktories K (1996) Inactivation of Ras by *Clostridium sordellii* lethal toxin-catalyzed glucosylation. J Biol Chem 271:10149–10153

Just I, Selzer J, Jung M, van Damme J, Vandekerckhove J, Aktories K (1995a) Rho-ADP-ribosylating exoenzyme from *Bacillus cereus*—purification, characterization and identification of the NAD-binding site. Biochemistry 34:334–340

Just I, Selzer J, Wilm M, Von Eichel-Streiber C, Mann M, Aktories K (1995b) Glucosylation of Rho proteins by *Clostridium difficile* toxin B. Nature 375:500–503

Just I, Wilm M, Selzer J, Rex G, Von Eichel-Streiber C, Mann M, Aktories K (1995c) The enterotoxin from *Clostridium difficile* (ToxA) monoglucosylates the Rho proteins. J Biol Chem 270:13932–13936

Kelly C P, LaMont J T (1998) *Clostridium difficile* infection. Annu Rev Med 49:375–390

Keusch J, Manzella S M, Nyame K A, Cummings R D, Baenziger J U (2000) Cloning of Gb_3 synthase, the key enzyme in globo-series glycosphingolipid synthesis, predicts a family of α1,4 glycosyltransferases conserved in plants, insects and mammals. J Biol Chem 275 (in press)

Koch G, Norgauer J, Aktories K (1994) ADP-ribosylation of Rho by *Clostridium limosum* exoenzyme affects basal but not N-formyl-peptide-stimulated actin polymerization in human myeloid leukaemic (HL60) cells. Biochem J 299:775–779

Kozma R, Ahmed S, Best A, Lim L (1995) The Ras-related protein Cdc42Hs and bradykinin promote formation of peripheral actin microspikes and filopodia in Swiss 3T3 fibroblasts. Mol Cell Biol 15:1942–1952

Krivan H C, Clark G F, Smith D F, Wilkins T D (1986) Cell surface binding site for *Clostridium difficile* enterotoxin: evidence for a glycoconjugate containing the sequence Galα1-3Galβ1-4GlcNAc. Infect Immun 53:573–581

Kuribara H, Tago K, Yokozeki T, Sasaki T, Takai Y, Morii N, Narumiya S, Katada T, Kanaho Y (1995) Synergistic activation of rat brain phospholipase D by ADP-ribosylation factor and *rho*A p21, and its inhibition by *Clostridium botulinum* C3 exoenzyme. J Biol Chem 270:25667–25671

Lamaze C, Chuang T H, Terlecky L J, Bokoch G M, Schmid S L (1996) Regulation of receptor-mediated endocytosis by Rho and Rac. Nature 382:177–179

Lang P, Guizani L, Vitté-Mony I, Stancou R, Dorseuil O, Gacon G, Bertoglio J (1992) ADP-ribosylation of the *ras*-related, GTP-binding protein RhoA inhibits lymphocyte-mediated cytotoxicity. J Biol Chem 267:11677–11680

Larson H E, Price A B, Honour P, Borriello S P (1978) *Clostridium difficile* and the aetiology of pseudomembranous colitis. Lancet 20:1063–1066

Larson H E, Proce A B (1977) Pseudomembranous colitis; presence of clostridial toxin. Lancet II:1312–1314

Laudanna C, Campbell J J, Butcher E C (1996) Role of Rho in chemoattractant-activated leukocyte adhesion through integrins. Science 271:981–983

Laudanna C, Campbell J J, Butcher E C (1997) Elevation of intracellular cAMP inhibits RhoA activation and integrin-dependent leukocyte adhesion induced by chemoattractants. J Biol Chem 272:24141–24144

Linseman D A, Hofmann F, Fisher S K (2000) A role for the small molecular weight GT-Pases, Rho and Cdc42, in muscarinic receptor signaling to focal adhesion kinase. J Neurochem 74:2010–2020

Liu T S, Musch M W, Sugi K, Walsh-Reitz M M, Ropeleski M J, Hendrickson B A, Pothoulakis C, LaMont J T, Chang e B (2003) Protective role of HSP72 against *Clostridium difficile* toxin A-induced intestinal epithelial cell dysfunction. Am J Physiol Cell Physiol 284:C1073–C1082

Lyerly D M, Krivan H C, Wilkins T D (1988) Clostridium difficile: its disease and toxins. Clin Microbiol Rev 1:1–18

Mackay D J G, Esch F, Furthmayr H, Hall A (1997) Rho- and Rac-dependent assembly of focal adhesion complexes and actin filaments in permeabilized fibroblasts: an essential role for Ezrin/Radixin/Moesin proteins. J Cell Biol 138:927–938

Mackay D J G, Hall A (1998) Rho GTPases. J Biol Chem 273:20685–20688

Mackay D J G, Nobes C D, Hall A (1995) The Rho's progress: A potential role during neuritogenesis for the Rho family of GTPases. Trends Neurosci 18:496–501

Maddala R, Deng P F, Costello J M, Wawrousek E F, Zigler J S, Rao V P (2004) Impaired cytoskeletal organization and membrane integrity in lens fibers of a Rho GTPase functional knockout transgenic mouse. Lab Invest 84:679–692

Madden J C, Ruiz N, Caparon M (2001) Cytolysin-mediated translocation (CMT): a functional equivalent type III secretion in gram-positive bacteria. Cell 104:143–152

Mahida Y R, Makh S, Hyde S, Gray T, Borriello S P (1996) Effect of *Clostridium difficile* toxin A on human intestinal epithelial cells: Induction of interleukin 8 production and apoptosis after cell detachment. Gut 38:337–347

Malcolm K C, Elliott C M, Exton J H (1996) Evidence for Rho-mediated agonist stimulation of phospholipase D in Rat1 fibroblasts—Effects of *Clostridium botulinum* C3 exoenzyme. J Biol Chem 271:13135–13139

Malorni W, Fiorentini C, Paradisi S, Giuliano M, Mastrantonio P, Donelli G (1990) Surface blebbing and cytoskeletal changes induced in vitro by toxin B form. Exp Mol Pathol 52:340–356

Mao J, Yuan H, Xie W, Simon M I, Wu D (1998a) Specific involvement of G proteins in regulation of serum response factor-mediated gene transcription by different receptors. J Biol Chem 273:27118–27123

Mao J, Yuan H, Xie W, Wu D (1998b) Guanine nucleotide exchange factor GEF115 specifically mediates activation of Rho and serum response factor by the G protein α subunit Gα13. Proc Natl Acad Sci USA 95:12973–12976

Martinez R D, Wilkins T D (1988) Purification and characterization of *Clostridium sordellii* hemorrhagic toxin and cross-reactivity with *Clostridium difficile* toxin A (enterotoxin). Infect Immun 56:1215–1221

Martinez R D, Wilkins T D (1992) Comparison of *Clostridium sordellii* toxins HT and LT with toxins A and B of *C. difficile*. J Med Microbiol 36:30–36

Masuda M, Betancourt L, Matsuzawa T, Kashimoto T, Takao T, Shimonishi Y, Horiguchi Y (2000) Activation of Rho through a cross-link with polyamines catalyzed by *Bordetella* dermonecrotizing toxin. EMBO J 19:521–530

Meacci E, Vasta V, Moorman J P, Bobak D A, Bruni P, Moss J, Vaughan M (1999) Effect of Rho and ADP-ribosylation factor GTPases on phospholipase D activity in intact human adenocarcinoma A549 cells. J Biol Chem 274:18605–18612

Mehlig M, Moos M, Braun V, Kalt B, Mahony D E, Von Eichel-Streiber C (2001) Variant toxin B and a functional toxin A produced by *Clostridium difficile* C34. FEMS Microbiol Lett 198:171–176

Ménétrey J, Flatau G, Stura E A, Charbonnier J-B, Gas F, Teulon J-M, Le Du M-H, Boquet P, Ménez A (2002) NAD binding induces conformational changes in Rho ADP-ribosylating *Clostridium botulinum* C3 exoenzyme. J Biol Chem 277:30950–30957

Meyer D K, Olenik C, Hofmann F, Barth H, Leemhuis J, Brünig I, Aktories K, Nörenberg W (2000) Regulation of somatodendritic $GABA_A$ receptor channels in rat hippocampal neurons: Evidence for a role of the small GTPase Rac1. J Neurosci 20:6743–6751

Moore R, Pothoulakis C, LaMont J T, Carlson S, Madara J L (1990) *C. difficile* toxin A increases intestinal permeability and induces Cl^-. Am J Physiol Gastrointest Liver Physiol 259:G165–G172

Moss J, Zolkiewska A, Okazaki I (1997) ADP-ribosylarginine hydrolases and ADP-ribosyltransferases—Partners in ADP-ribosylation cycles. Adv Exp Med Biol 419:25–33

Negishi M, Dong J, Darden T A, Pedersen L G, Pedersen L C (2003) Glucosaminylglycan biosynthesis: what we can learn from the X-ray crystal structures of glycosyltransferases GlcAT1 and EXTL2. Biochem Biophys Res Commun 303:393–398

Nemoto E, Yu Y J, Dennert G (1996) Cell surface ADP-Ribosyltransferase regulates lymphocyte function-associated molecule-1 (LFA-1) function in T cells. J Immunol 157:3341–3349

Nobes C D, Hall A (1995) Rho, Rac, and Cdc42 GTPases regulate the assembly of multimolecular focal complexes associated with actin stress fibers, lamellipodia, and filopodia. Cell 81:53–62

Nusrat A, Giry M, Turner J R, Colgan S P, Parkos C A, Carnes D, Lemichez E, Boquet P, Madara J L (1995) Rho protein regulates tight junctions and perijunctional actin organization in polarized epithelia. Proc Natl Acad Sci USA 92:10629–10633

Nusrat A, Von Eichel-Streiber C, Turner J R, Verkade P, Madara J L, Parkos C A (2001) *Clostridium difficile* toxins disrupt epithelial barrier function by altering membrane microdomain localization of tight junction proteins. Infect Immun 69:1329–1336

Oksche A, Nakov R, Habermann E (1992) Morphological and biochemical study of cytoskeletal changes in cultured cells after extracellular application of *Clostridium novyi* α-toxin. Infect Immun 60:3002–3006

Olson M F, Paterson H F, Marshall C J (1998) Signals from Ras and Rho GTPases interact to regulate expression of $p21^{Waf1/Cip1}$. Nature 394:295–299

Ottlinger M E, Lin S (1988) *Clostridium difficile* toxin B induces reorganization of actin, vinculin, and talin in cultured cells. Exp Cell Res 174:215–229

Paterson H F, Self A J, Garrett M D, Just I, Aktories K, Hall A (1990) Microinjection of recombinant $p21^{rho}$ induces rapid changes in cell morphology. J Cell Biol 111:1001–1007

Pfeifer G, Schirmer J, Leemhuis J, Busch C, Meyer D K, Aktories K, Barth H (2003) Cellular uptake of *Clostridium difficile* toxin B: translocation of the N-terminal catalytic domain into the cytosol of eukaryotic cells. J Biol Chem 278:44535–44541

Popoff M R, Chaves O E, Lemichez E, Von Eichel-Streiber C, Thelestam M, Chardin P, Cussac D, Chavrier P, Flatau G, Giry M, Gunzburg J, Boquet P (1996) Ras, Rap, and Rac small GTP-binding proteins are targets for *Clostridium sordellii* lethal toxin glucosylation. J Biol Chem 271:10217–10224

Pothoulakis C, Castagliuolo I, LaMont J T, Jaffer A, O'Keane J C, Snider R M, Leeman S E (1994) CP-96,345, a substance P antagonist, inhibits rat intestinal responses to *Clostridium difficile* toxin A but not cholera toxin. Proc Natl Acad Sci USA 91:947–951

Pothoulakis C, Gilbert R J, Cladaras C, Castagliuolo I, Semenza G, Hitti Y, Montcrief J S, Linevsky J, Kelly C P, Nikulasson S, Desai H P, Wilkins T D, LaMont J T (1996) Rabbit sucrase-isomaltase contains a functional intestinal receptor for *Clostridium difficile* toxin A. J Clin Invest 98:641–649

Pothoulakis C, LaMont J T (2001) Microbes and microbial toxins: paradigms for microbial-mucosal interactions. II. The integrated response of the intestine to *Clostridium difficile* toxins. Am J Physiol Gastrointest Liver Physiol 280:G178–G183

Pothoulakis C, Sullivan R, Melnick D A, Triadafilopoulos G, Gadenne A-S, Meshulam T, LaMont J T (1988) *Clostridium difficile* toxin A stimulates intracellular calcium release and chemotactic response in human granulocytes. J Clin Invest 81:1741–1745

Prepens U, Just I, Von Eichel-Streiber C, Aktories K (1996) Inhibition of FcεRI-mediated activation of rat basophilic leukemia cells by *Clostridium difficile* toxin B (monoglucosyltransferase). J Biol Chem 271:7324–7329

Qa'Dan M, Spyres L M, Ballard J D (2000) pH-induced conformational changes in *Clostridium difficile* toxin B. Infect Immun 68:2470–2474

Reif K, Cantrell D A (1998) Networking Rho family GTPases in lymphocytes. Immunity 8:395–401

Ren X-D, Bokoch G M, Traynor-Kaplan A, Jenkins G H, Anderson R A, Schwartz M A (1996) Physical association of the small GTPase Rho with a 68-kDa phosphatidylinositol 4-phosphate 5-kinase in swiss 3T3 cells. Mol Biol Cell 7:435–442

Ridley A J, Hall A (1992) The small GTP-binding protein rho regulates the assembly of focal adhesions and actin stress fibers in response to growth factors. Cell 70:389–399

Ridley A J, Hall A (1994) Signal transduction pathways regulating Rho-mediated stress fibre formation: Requirement for a tyrosine kinase. EMBO J 13:2600–2610

Ridley A J, Paterson H F, Johnston C L, Diekmann D, Hall A (1992) The small GTP-binding protein rac regulates growth factor- induced membrane ruffling. Cell 70:401–410

Riegler M, Sedivy R, Pothoulakis C, Hamilton G, Zacheri J, Bischof G, Cosentini E, Feil W, Schiessel R, LaMont J T, Wenzl E (1995) *Clostridium difficile* toxin B is more potent than toxin A in damaging human colonic epithelium in vitro. J Clin Invest 95:2004–2011

Rifkin G D, Fekety F R, Silva J, Sack R B (1977) Antibiotic-induced colitis. Implication of a toxin neutralised by *Clostridium sordellii* antitoxin. Lancet II:1103–1106

Rolfe R D, Song W (1993) Purification of a functional receptor for *Clostridium difficile* toxin A from intestinal brush border membranes of infant hamsters. Clin Infect Dis 16:219–227

Rolfe R D, Song W (1995) Immunoglobulin and non-immunoglobulin components of human milk inhibit *Clostridium difficile* toxin A-receptor binding. J Med Microbiol 42:10–19

Rubin E J, Gill D M, Boquet P, Popoff M R (1988) Functional modification of a 21-Kilodalton G protein when ADP-ribosylated by exoenzyme C3 of *Clostridium botulinum*. Mol Cell Biol 8:418–426

Rupnik M, Avesani V, Janc M, Von Eichel-Streiber C, Delmée M (1998) A novel toxinotyping scheme and correlation of toxinotypes with serogroups of *Clostridium difficile* isolates. J Clin Microbiol 36:2240–2247

Rupnik M, Braun V, Soehn F, Janc M, Hofstetter M, Laufenberg-Feldmann R, Von Eichel-Streiber C (1997) Characterization of polymorphisms in the toxin A and B genes of *Clostridium difficile*. FEMS Microbiol Lett 148:197–202

Sah V P, Hoshijima M, Chien K R, Brown J H (1996) Rho is required for $G\alpha_q$ and α_1-adrenergic receptor signaling in cardiomyocytes—Dissociation of Ras and Rho pathways. J Biol Chem 271:31185–31190
Saito Y, Nemoto Y, Ishizaki T, Watanabe N, Morii N, Narumiya S (1995) Identification of Glu^{173} as the critical amino acid residue for the ADP-ribosyltransferase activity of *Clostridium botulinum* C3 exoenzyme. FEBS Lett 371:105–109
Santos M F, McCormack S A, Guo Z, Okolicany J, Zheng Y, Johnson L R, Tigyi G (1997) Rho proteins play a critical role in cell migration during the early phase of mucosal restitution. J Clin Invest 100:216–225
Sauerborn M, Leukel P, Von Eichel-Streiber C (1997) The C-terminal ligand-binding domain of *Clostridium difficile* toxin A (TcdA) abrogates TcdA-specific binding to cells and prevents mouse lethality. FEMS Microbiol Lett 155:45–54
Sauerborn M, Von Eichel-Streiber C (1990) Nucleotide sequence of *Clostridium difficile* toxin A. Nucleic Acids Res 18:1629–1630
Savidge T C, Pan W-H, Newman P, O'Brien M, Anton P M, Pothoulakis C (2003) *Clostridium difficile* toxin B is an inflammatory enterotoxin in human intestine. Gastroenterology 125:413–420
Schmalzing G, Richter H P, Hansen A, Schwarz W, Just I, Aktories K (1995) Involvement of the GTP binding protein Rho in constitutive endocytosis in *Xenopus laevis* oocytes. J Cell Biol 130:1319–1332
Schmidt G, Sehr P, Wilm M, Selzer J, Mann M, Aktories K (1997) Gln63 of Rho is deamidated by *Escherichia coli* cytotoxic necrotizing factor 1. Nature 387:725–729
Schmidt M, Rümenapp U, Bienek C, Keller J, Von Eichel-Streiber C, Jakobs K H (1996) Inhibition of receptor signaling to phospholipase D by *Clostridium difficile* toxin B—Role of Rho proteins. J Biol Chem 271:2422–2426
Schmidt M, Voss M, Weernink P A, Wetzel J, Amano M, Kaibuchi K, Jakobs K H (1999) A role for Rho-kinase in Rho-controlled phospholipase D stimulation by the m3 muscarinic acetylcholine receptor. J Biol Chem 274:14648–14654
Sehr P, Joseph G, Genth H, Just I, Pick E, Aktories K (1998) Glucosylation and ADP-ribosylation of Rho proteins—Effects on nucleotide binding, GTPase activity, and effector-coupling. Biochemistry 37:5296–5304
Sekine A, Fujiwara M, Narumiya S (1989) Asparagine residue in the rho gene product is the modification site for botulinum ADP-ribosyltransferase. J Biol Chem 264:8602–8605
Selzer J, Hofmann F, Rex G, Wilm M, Mann M, Just I, Aktories K (1996) *Clostridium novyi* α-toxin-catalyzed incorporation of GlcNAc into Rho subfamily proteins. J Biol Chem 271:25173–25177
Servant G, Weiner O D, Herzmark P, Balla T, Sedat J W, Bourne H R (2000) Polarization of chemoattractant receptor signaling during neutrophil chemotaxis. Science 287:1037–1040
Shao F, Merritt P M, Bao Z, Innes R W, Dixon J E (2002) A *Yersinia* effector and a *Pseudomonas* avirulence protein define a family of cysteine proteases functioning in bacterial pathogenesis. Cell 109:575–588
Soehn F, Wagenknecht-Wiesner A, Leukel P, Kohl M, Weidmann M, Von Eichel-Streiber C, Braun V (1998) Genetic rearrangements in the pathogenicity locus of *Clostridium difficile* strain 8864: implications for transcription, expression and enzymatic activity of toxins A and B. Mol Gen Genet 258:222–232
Stam J C, Michiels F, Van der Kammen R A, Moolenaar W H, Collard J G (1998) Invasion of T-lymphoma cells: cooperation between Rho family GTPases and lysophospholipid receptor signaling. EMBO J 17:4066–4074

Stasia M-J, Jouan A, Bourmeyster N, Boquet P, Vignais P V (1991) ADP-ribosylation of a small size GTP-binding protein in bovine neutrophils by the C3 exoenzyme of *Clostridium botulinum* and effect on the cell motility. Biochem Biophys Res Commun 180:615–622

Subauste M C, Von Herrath M, Benard V, Chamberlain C E, Chuang T H, Chu K, Bokoch G M, Hahn K M (2000) Rho family proteins modulate rapid apoptosis induced by cytotoxic T lymphocytes and Fas. J Biol Chem 275:9725–9733

Sugai M, Enomoto T, Hashimoto K, Matsumoto K, Matsuo Y, Ohgai H, Hong Y-M, Inoue S, Yoshikawa K, Suginaka H (1990) A novel epidermal cell differentiation inhibitor (EDIN): Purification and characterization from *Staphylococcus aureus.* Biochem Biophys Res Commun 173:92–98

Takai Y, Sasaki T, Matozaki T (2001) Small GTP-binding proteins. Physiol Rev 81:153–208

Taylor N S, Thorne G M, Bartlett J G (1981) Comparison of two toxins produced by *Clostridium difficile.* Infect Immun 34,No.3:1036–1043

Teneberg S, Lönnroth I, López J F T, Galili U, Halvarsson M Ö, Ångström J, Karlsson K A (1996) Molecular mimicry in the recognition of glycosphingolipids by Galα3Galβ4GlcNAcβ-binding *Clostridium difficile* toxin A, human natural anti α-galactosyl IgG and the monoclonal antibody Gal-13: Characterization of a binding-active human glycosphingolipid, non-identical with the animal receptor. Glycobiology 6:599–609

Thelestam M, Chaves-Olarte E (2000) Cytotoxic effects of the *Clostridium difficile* toxins. Curr Top Microbiol Immunol 250:85–96-

Tigyi G, Fischer D J, Sebök A, Marshall F, Dyer D L, Miledi R (1996) Lysophosphatidic acid-induced neurite retraction in PC12 cells: Neurite-protective effects of cyclic AMP signaling. J Neurochem 66:549–558

Tucker K D, Wilkins T D (1991) Toxin A of *Clostridium difficile* binds to the human carbohydrate antigens I, X, and Y. Infect Immun 59:73–78

Van Aelst L, D'Souza-Schorey C (1997) Rho GTPases and signaling networks. Genes Dev 11:2295–2322

Verschueren H, De Baetselier P, De Braekeleer J, Dewit J, Aktories K, Just I (1997) ADP-ribosylation of Rho-proteins with botulinum C3 exoenzyme inhibits invasion and shape changes of T-lymphoma cells. Eur J Cell Biol 73:182–187

Vetter I R, Hofmann F, Wohlgemuth S, Herrmann C, Just I (2000) Structural consequences of mono-glucosylation of Ha-Ras by *Clostridium sordellii* lethal toxin. J Mol Biol 301:1091–1095

Vögler O, Krummenerl P, Schmidt M, Jakobs K H, van Koppen C J (1999) RhoA-sensitive trafficking of muscarinic acetylcholine receptors. J Pharmacol Exp Ther 288:36–42

Von Eichel-Streiber C, Boquet P, Sauerborn M, Thelestam M (1996) Large clostridial cytotoxins—a family of glycosyltransferases modifying small GTP-binding proteins. Trends Microbiol 4:375–382

Von Eichel-Streiber C, Laufenberg-Feldmann R, Sartingen S, Schulze J, Sauerborn M (1992) Comparative sequence analysis of the *Clostridium difficile* toxins A and B. Mol Gen Genet 233:260–268

von Pawel-Rammingen U, Telepnev M V, Schmidt G, Aktories K, Wolf-Watz H, Rosqvist R (2000) GAP activity of the *Yersinia* YopE cytotoxin specifically targets the Rho pathway: a mechanism for disruption of actin microfilament structure. Mol Microbiol 36:737–748

Vouret-Craviari V, Boquet P, Pouysségur J, Van Obberghen-Schilling E (1998) Regulation of the actin cytoskeleton by thrombin in human endothelial cells: Role of Rho proteins in endothelial barrier function. Mol Biol Cell 9:2639–2653

Wahl S, Barth H, Ciossek T, Aktories K, Mueller B K (2000) Ephrin-A5 induces collapse of growth cones by activating Rho and Rho kinase. J Cell Biol 149:263–270

Warny M, Keates A C, Keates S, Castagliuolo I, Zacks J K, Aboudola S, Qamar A, Pothoulakis C, LaMont J T, Kelly C P (2000) p38 MAP kinase activation by *Clostridium difficile* toxin A mediates monocyte necrosis, IL-8 production, and enteritis. J Clin Invest 105:1147–1156

Watanabe N, Madaule P, Reid T, Ishizaki T, Watanabe G, Kakizuka A, Saito Y, Nakao K, Jockusch B M, Naumiya S (1997) p140mDia, a mammalian homolog of *Drosophila* diaphanous, is a target protein fro Rho small GTPase and is a ligand for profilin. EMBO J 16:3044–3056

Weernick P A O, Schulte P, Guo Y, Wetzel J, Amano M, Kaibuchi K, Haverland S, Voß M, Schmidt M, Mayr G W, Jakobs K H (2000) Stimulation of phosphatidylinositol-4-phosphate 5-kinase by Rho-kinase. J Biol Chem 275:10168–10174

Wei Y, Zhang Y, Derewenda U, Liu X, Minor W, Nakamoto R K, Somlyo A V, Somlyo A P, Derewenda Z S (1997) Crystal structure of RhoA-GDP and its functional implications. Nat Struct Biol 4:699–703

Wennerberg K, Der C J (2004) Rho-family GTPases: it's not only Rac and Rho (and I like it). J Cell Sci 117:1301–1312

Wershil B K, Castagliuolo I, Pothoulakis C (1998) Direct evidence of mast cell involvement in *Clostridium difficile* toxin A-induced enteritis in mice. Gastroenterology 114:956–964

Wiegers W, Just I, Müller H, Hellwig A, Traub P, Aktories K (1991) Alteration of the cytoskeleton of mammalian cells cultured in vitro by *Clostridium botulinum* C2 toxin and C3 ADP-ribosyltransferase. Eur J Cell Biol 54:237–245

Wiggins C A R, Munro S (1998) Activity of the yeast *MNN1* α-1,3-mannosyltransferase requires a motif conserved in many other families of glycosyltransferases. Proc Natl Acad Sci USA 95:7945–7950

Wilde C, Barth H, Sehr P, Han L, Schmidt M, Just I, Aktories K (2002) Interaction of the Rho-ADP-ribosylating C3 exoenzyme with RalA. J Biol Chem 277:14771–14776

Wilde C, Chhatwal G S, Aktories K (2001a) C3stau, a new member of the family of C3-like ADP-ribosyltransferases. Trends Microbiol 10:5–7

Wilde C, Chhatwal G S, Schmalzing G, Aktories K, Just I (2001b) A novel C3-like ADP-ribosyltransferase from *Staphylococcus aureus* modifying RhoE and Rnd3. J Biol Chem 276:9537–9542

Wilkins T D (1987) Role of *Clostridium difficile* toxins in disease. Gastroenterology 93:389–391

Wojciak-Stothard B, Entwistle A, Garg R, Ridley A J (1998) Regulation of TNF-α-induced reorganization of the actin cytoskeleton and cell-cell junctions by Rho, Rac, and Cdc42 in human endothelial cells. J Cell Physiol 176:150–165

Yamaguchi T, Hayashi T, Takami H, Ohnishi M, Murata T, Nakayama K, Asakawa K, Ohara M, Komatsuzawa H, Sugai M (2001) Complete nucleotide sequence of a *Staphylococcus aureus* exfoliative toxin B plasmid and identification of a novel ADP-ribosyltransferase, EDIN-C. Infect Immun 69:7760–7771

CTMI (2005) 291:147–166

The Type III Cytotoxins of *Yersinia* and *Pseudomonas aeruginosa* That Modulate the Actin Cytoskeleton

M. R. Baldwin · J. T. Barbieri (✉)

Microbiology and Molecular Genetics, Medical College of Wisconsin,
8701 Watertown Plank Road, Milwaukee WI, 53225, USA
jtb01@mcw.edu

Abstract Initial studies of how bacterial toxins modulate the actin cytoskeleton have focused primarily on the mode of action of these toxins. More recently, studies have addressed the molecular interactions of these toxins with host cell signaling pathways and how toxins modulate cellular physiology. Although each individual toxin has a unique mode of action, general themes have started to emerge between bacterial pathogens. During the course of an infection, many pathogenic bacteria produce toxins that target the actin cytoskeleton and its regulatory proteins. Toxins can either act as positive regulators promoting the assembly of filamentous actin structures or, alternatively, as negative regulators promoting actin filament disassembly. Modulation of the actin cytoskeleton facilitates various infectious processes critical for the success of the pathogen. Intracellular bacteria such as *Salmonella typhimurium* utilize toxins to promote both assembly and disassembly of the actin cytoskeleton during the infection process. Temporal regulation of toxin activities results in internalization of the bacterium by epithelial cells into specialized vacuoles permissive for growth. In contrast, *Yersinia* utilizes actin modulating toxins to block internalization by professional antigen-presenting cells such as macrophages and dendritic cells. Modulation of the immune response through the production of actin-regulating toxins appears to be a common approach adopted by several extracellular pathogens. Thus the repertoire of actin-modifying toxins produced by various species is specifically tailored to facilitate the lifestyle of the pathogen. The presence of multiple toxins that modulate the activation state of actin shows the importance of interfering with the cytoskeleton to neutralize the host's innate immune system for the survival and growth of *Yersinia* and *P. aeruginosa.*

1 Introduction

Bacterial pathogens often utilize toxins to modulate the host response to infection. *Yersinia* and *Pseudomonas aeruginosa* are extracellular pathogens that produce type III cytotoxins, where the cytotoxin is delivered directly into host cells by a contact-mediated apparatus that is a component of the bacterium. Several *Yersinia* and *P. aeruginosa* type III cytotoxins reorganize the host's actin cytoskeleton, disrupting the host innate immune response. The type III cytotoxins differ from "classic" bacterial exotoxins that are organized into three domains; a catalytic A domain that contains the enzymatic action of the toxin and a B domain that comprises two functions, a binding domain that binds to receptors on the surface of sensitive cells to mediate the specificity of intoxication and a translocation domain that delivers the A domain across the host cell membrane. This chapter provides an overview of the *Yersinia* and *P. aeruginosa* type III cytotoxins that modulate the host actin cytoskeleton.

2 Molecular Pathogenesis of *P. aeruginosa* and *Yersinia*

P. aeruginosa is a ubiquitous, opportunistic pathogen of compromised patients, including individuals with severe burn wounds, eye complications, neutropenia, and cystic fibrosis. *P. aeruginosa* pathogenesis is facilitated by its resistance to antibiotics [14, 39]. The genome sequence of strain PA01 [65] has facilitated studies on virulence determinants, which are either on the cell surface or secreted. Cell surface components allow *P. aeruginosa* to colonize the site of infection, subvert the immune system, and replicate within the host. These factors include endotoxin [25], fimbriae [58], flagella [12, 63], and alginate [18, 52, 69], a polysaccharide capsule that is evident in lung isolates. *P. aeruginosa* secretes several virulence factors, including proteases [38, 51], phospholipases [70], and siderophores [40, 71]. *P. aeruginosa* produces one ADP-ribosylating exotoxin, exotoxin A [27]. Exotoxin A is a single-chain protein with defined AB structure-function properties. The amino terminus encodes the receptor binding and translocation domains (B), whereas the carboxyl terminus comprises the ADP ribosyltransferase domain (A). Exotoxin A binds to the LDL-like receptor, enters cells via receptor-mediated endocytosis [15], and is retrograde transported to the endoplasmic reticulum, where the A domain is translocated into the cytosol. Exotoxin A ADP-ribosylates eukaryotic elongation factor-2, which inhibits protein synthesis and causes cell death [27]. Type III cytotoxins of *P. aeruginosa* [75] are a new class of virulence factors that contribute to the subversion of the innate immune response. *P. aeruginosa* type III secreted cytotoxins include ExoU, ExoY, ExoS, and ExoT. ExoU, a potent cytotoxin that possesses phospholipase activity [13], and ExoY, an adenylate cyclase, are activated by mammalian proteins [76]. ExoS and ExoT are bifunctional cytotoxins that reorganize the actin cytoskeleton via a RhoGAP activity and ADP-ribosylate host proteins; they will be discussed in this review along with ExoY. *Yesinia pestis* is the etiologic agent of plague and a category A agent. *Y. pestis* shares numerous features with *Y. enterocolitica* and *Y. pseudotuberculosis*, including a tropism for lymphoid tissues and the ability to resist the host innate immune system. Modulation of the host innate immune system is due to the presence of a large virulence plasmid, which encodes a type III secretion apparatus and type-III cytotoxins (termed Yops; *Yersinia* outer membrane proteins). The Yops were initially described [41] as plasmid-encoded proteins secreted by *Yersinia* on incubation at 37°C in the absence of Ca^{2+}. The function of effector proteins (now termed type III cytotoxins) were identified on directed introduction into mammalian cells either by microinjection [49] or by the bacterial type III secretion system [50, 62]. Type III delivered cytotoxins are now recognized as a common virulence mechanism for Gram-negative bacterial pathogens to subvert cells of the innate immune system [26]. The *Yersinia* type III cytotoxins YopH, YopT, and YopE directly modulate the host actin cytoskeleton and will be discussed in this review.

3 Secretion of Type III Cytotoxins

The type-III secretion systems of *Yersinia* and *P. aeruginosa* are similar and comprise two components, which are required for secretion of type III cytotoxins out of the bacterium and for translocation across the plasma membrane into the mammalian cell. The *Yersinia* type III apparatus is composed of 30 *ysc* (*Yersinia* secretion) genes and is a product of gene duplication of the flagellum operon. The functions of individual components of the type III apparatus are known. YscC is essential for the secretion of Yops across the outer membrane of *Yersinia* and is a member of the family of outer membrane "secretin" proteins that function in macromolecular transport. YscC forms an oligomeric complex in the outer membrane of *Yersinia* with an apparent internal pore [32]. Secretion of type III cytotoxins also requires YscN, which possesses an ATP-binding motif [73]. Mutations within this ATP binding motif interfere with Yop secretion, implicating a need for energy in the secretion process.

4 Translocation of Type III Cytotoxins into Mammalian Cells: LcrV, YopB, and YopD

LcrV regulates type III secretion as a low-calcium response that control regulation of Yop expression [6]. Straley and coworkers showed that LcrV was necessary for full induction of the low-calcium response [61]. LcrV is involved in the release of YopB and YopD from the bacterial cytoplasm and may stimulate secretion through direct protein-protein interactions [53]. Straley and coworkers implicated a *ysc*-independent translocation of LcrV into mammalian cells, LcrV was localized to the bacterial cytoplasm, with some present in the extracellular medium [10, 36].

Yersinia utilize YopB and YopD to deliver type III cytotoxins into mammalian cells. Early indications for a role of YopB in type III translocation included the observation that nonpolar mutations within YopB did not interfere with *Yersinia* secretion of type III cytotoxins into the culture medium but prevented their intracellular delivery into the eukaryotic cell [24]. Initial recognition of the pore-forming properties of YopB/YopD were made by Neyt and Cornelis [42], who subsequently showed that YopB/YopD were required for type III-mediated pore formation in macrophages [42]. The pore generated by YopB/YopD allows diffusion of low-molecular-weight agents (443 and 623 Da, but not 1,490 Da), indicating that protein translocation through the pore would require at least partial protein unfolding. PcrV, PopB, and PopD are the *Pseudomonas* homologs of the *Yersinia* proteins LcrV, YopB, and YopD, respectively, and possess similar functional properties [16, 55]. Continued studies should provide a detailed understanding of the molecular regulation of type III secretion and translocation.

5
The Rho GTPase Cycle and the Organization of the Actin Cytoskeleton

Activation or inhibition of the actin cytoskeleton by bacterial toxins provides the pathogen with unique advantages within the host. Cytoskeleton modification allows the pathogen to resist phagocytosis by professional phagocytes or to invade specialized cells, such as epithelial and endothelial cells. Bacterial toxins target actin and the monomeric GTP-binding Rho proteins that modulate the state of the actin cytoskeleton. The state of actin fluctuates between soluble monomeric actin (G-actin) and polymerized actin (F-actin), especially during phagocytosis [8]. F-actin is polymerized in a polar process that involves slow actin growth at one termini and fast actin growth at the other termini. Actin-binding proteins can cap, bundle, and sever F-actin, which defines polymerization status. Arp2/3 is an actin-binding protein that forms a protein complex to recruit profilin-actin-ATP to the F-actin terminus. Profilin is an exchange factor that binds actin-ADP and stimulates nucleotide exchange to generate G-actin-ATP [54]. G-actin- ATP has a decreased critical concentration for association to F-actin and binds the fast growing end of actin filaments. Barbed-end capping proteins can also bind actin filaments to inhibit actin polymerization. Within the actin filament F-actin-ATP is hydrolyzed to F-actin-ADP. F-actin-ADP is then released as an actin monomer from the slow-growing end of the actin filament. During actin depolymerization, cofilin stimulates the hydrolysis of actin-GTP to actin-GDP and can sever capped actin filaments, which expose the actin-ADP to stimulate filament depolymerization [77]. Actin nucleation and formation of stress fibers, filopodia, and lamellipodia are regulated by the Rho GTPases [47] Rho, Rac, and Cdc42. The Rho GTPases act as molecular switches to control signal transduction pathways by cycling between a GDP-bound, inactive form and a GTP-bound, active form. In their GTP-bound state, they interact with downstream host proteins to elicit a variety of intracellular responses, the best-characterized function of which is the regulation of actin dynamics. In Swiss 3T3 cells constitutively active (RhoA Gly14Val and Gln63Glu) and dominant negative, interfering forms (RhoA Thr19Asn) showed that Rho regulates the assembly of contractile, actin:myosin stress fibers, whereas Rac and Cdc42 regulate the polymerization of actin to form peripheral lamellipodia to modulate membrane ruffling and cellular locomotion and Cdc42 regulates filopodia to define the polarity of cellular motility [43]. The interaction of Rho GTPases with the plasma membrane is mediated by posttranslational lipid modification of the proteins. Rho GTPases undergo geranylgeranylation catalyzed by geranylgeranyl transferase type I (GGTase-I) at the Cys residue of the C-terminal CAAX motif. Recognition of the *CAAX* motif is determined by the last amino acid in the *CAAX* sequence, with leucine forming the substrate for GGTase-I. The Rho GTPase cycle is tightly regulated by three groups of proteins. Inactive Rho-GDP is sequestered and solubilized in the cytoplasm through complex formation with guanine nucleotide dissociation inhibitors (GDIs) [47]. ERM proteins and lipids stimulate GDI release from Rho, which unmasks the lipophilic isoprenyl group, facilitating the transfer of the Rho GTPase to the cell membrane. Release of GDI allows Rho to associate with guanine nucleotide exchange factors (GEFs), which stimulate conversion of Rho-GDP to Rho-GTP. The membrane-bound Rho-GTP associates with effector proteins to stimulate actin cytoskeleton reorganization. Direct downstream effectors of Rho GTPases include protein kinases and lipid kinases, such as Rho kinase and PIP kinase for Rho and PAK for Rac and Cdc42. Inactivation of Rho-GTP to Rho- GDP is stimulated by Rho specific GTPase-activating proteins (GAPs), which negatively regulate the switch by enhancing its intrinsic GTPase activity. Rho-GDP is extracted from the cell membrane by GDI to complete the cycle. Although the proteins involved in Rho GTPase cycling are known, the downstream events that stimulate actin cytoskeleton reorganization are not completely understood.

6 *P. aeruginosa* Type III Cytotoxins That Modulate the Actin Cytoskeleton: Rho GAP Domains of ExoS and ExoT

Exoenzyme S is a bifunctional type III cytotoxin that possesses two independent activities (Fig. 1). The N terminus comprises a Rho GAP activity, whereas the C terminus comprises an ADP-ribosyltransferase activity. Iglewski and coworkers discovered ExoS as a second ADP-ribosylating protein that was produced by *P. aeruginosa* [28]. ExoS ADP-ribosylated Ras and several related GTPases [9], with subsequent studies showing that ADP-ribosylation of Ras at Arg41 disrupted Ras interactions with its GEF [19]. The FAS-dependent ADP-ribosyltransferase domain was localized to the C terminus of ExoS [31]. Early studies by Fritz-Lindsten et al. suggested the presence of a second activity within ExoS, because an ADP-ribosyltransferase defective form of ExoS retained the ability to affect the mammalian cell cytoskeleton. Subsequent studies showed that the N terminus of ExoS possessed a domain that stimulated actin reorganization [45] as a Rho GAP activity [23]. ExoS Rho GAP was active on the three major classes of Rho GTPases (Rho, Rac, and Cdc42) in both in vitro assays and cultured cells [34]. Wurtele et al. [74] solved the crystal structure of the ExoS Rho GAP domain complexed to Rac1 (Fig. 1). The Rho GAP domain was composed of 130 amino acids and was organized in an α-helical conformation. The ExoS Rho GAP domain had no obvious structural

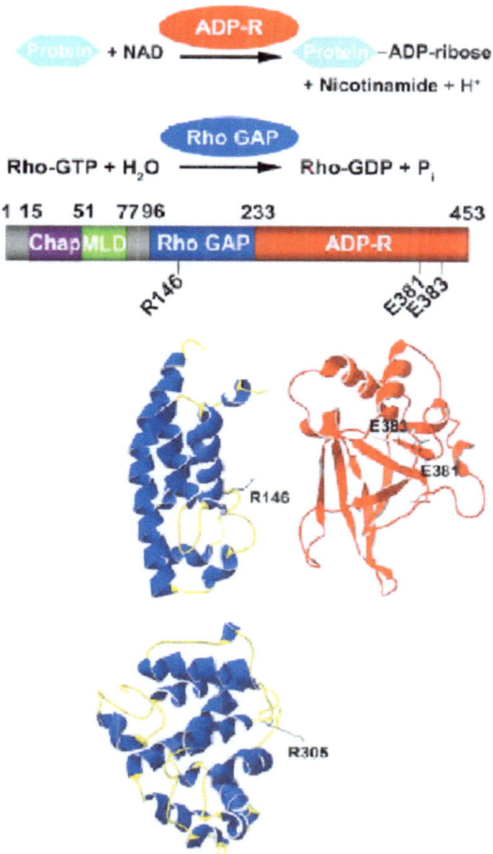

Fig. 1 Structural arrangement of the *Pseudomonas aeruginosa* effector ExoS. The *Pseudomonas* virulence factor ExoS is a bifunctional protein containing two unique enzyme activities. The N terminus contains domains involved in chaperone binding (*Chap*, residues 15–51) and membrane localization (*MLD*, residues 51–77). The central domain contains Rho GAP activity (residues 96–233) catalyzing conversion of the active GTP-bound form of RhoA to the inactive GDP-bound form. The position of the catalytic arginine 146 residue is highlighted in *bold*. The C-terminal ADP ribosyltransferase domain (*ADP-R*, residues 234–453) catalyzes the transfer of ADP-ribose from NAD to several protein substrates. The catalytic glutamate residues E381 and E383 are highlighted in *bold*. The crystal structures of the ExoS Rho GAP (*middle left panel*, shown in *blue*) and the mammalian Cdc42 GAP domain (*lower panel*, shown in *blue*) exhibit different folds to each other, which suggests that they are the products of convergent evolution. The ExoS catalytic Arg146 residue is contained on an α-helix, rather than a flexible loop as is the case for the mammalian Cdc42 GAP. A model of the ExoS ADP ribosyltransferase domain (*middle right panel*, shown in *red*) was constructed with Swiss PDB viewer by alignment of the primary sequences with the known structures of C3, VIP2, and iota toxins (ExoS Rho GAP from PDB: 1HE1, Cdc42 GAP domain from PDB: 1GRN)

homology, and thus no recognizable evolutionary relationship, with mammalian RhoGAPs. Although not sharing obvious homology to mammalian Rho GAPs, ExoS RhoGAP shares structural homology to the bacterial GAPs YopE and SptP [11, 64]. The three bacterial GAPs utilize an arginine finger to stabilize the transition state of the GTPase reaction. ExoS possesses a short hydrophobic region (residues 51–77, termed membrane localization domain, MLD) that targets the toxin to intracellular membranes in mammalian cells [44]. YopE was recently observed [35] to contain a short hydrophobic region (residues 54–75) that was necessary and sufficient for intracellular localization in mammalian cells. The YopE localization domain complemented the ExoS MLD for intracellular targeting in mammalian cells, suggesting conservation in function [33, 35]. ExoT of *P. aeruginosa* possesses 76% homology with ExoS, and, like ExoS, the N terminus of ExoT is a Rho GAP for RhoA, Rac1, and Cdc42 [20, 33] (Fig. 2). The Rho GAP domain of ExoT stimulates the reorganization of the actin cytoskeleton in cultured cells, inhibits internalization of *P. aeruginosa* by epithelial cells and macrophages [21], and inhibits epithelial wound repair [22]. This inhibition occurs primarily at the edge of the wound and results from actin cytoskeleton collapse and cell rounding and detachment. ExoT may allow *P. aeruginosa* to overwhelm the host innate defenses, such as an intact epithelial barrier, and evade phagocytosis.

7 ExoY Adenylate Cyclase

ExoY was discovered as a fourth type III cytotoxin of *P. aeruginosa* [76]. ExoY has homology to the active site regions of adenylate cyclase toxin of *B. pertussis* and edema factor of *B. anthracis*, including the ATP-binding motif. The specific activity of ExoY for the formation of cAMP is similar to the basal activity of adenylate cyclase toxin. Although both ExoY and adenylate cyclase toxin are stimulated by mammalian proteins, the activator of ExoY is not calmodulin, which activates adenylate cyclase toxin.

8 ExoT ADP-Ribosylates Crk Proteins: A Novel Mechanism of Antiphagocytosis

ExoT is a 457-amino acid bifunctional type III secreted cytotoxin of *P. aeruginosa* that contains an N-terminal Rho GAP domain and a C-terminal ADP-ribosylation domain (Fig. 2). Although the Rho GAP activity of ExoT re-

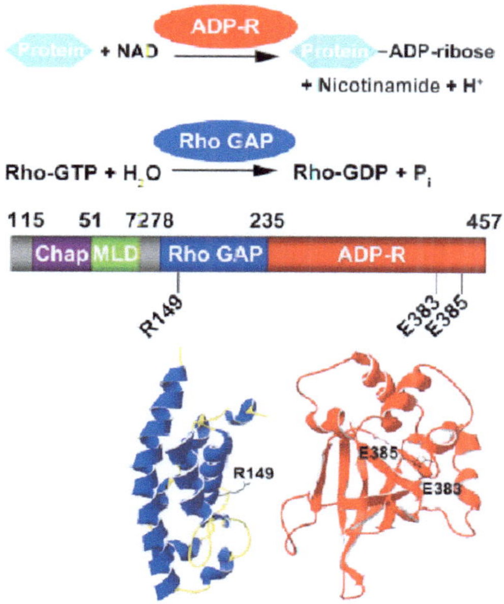

Fig. 2 Structural arrangement of the *Pseudomonas aeruginosa* effector ExoT. The *Pseudomonas* virulence factor ExoT is a bifunctional protein containing two unique enzyme activities. The N terminus contains domains involved in chaperone binding (Chap, residues 15–51) and membrane localization (*MLD*, residues 51–72). The central domain contains Rho GAP activity (residues 78–235) catalyzing conversion of the active GTP-bound form of RhoA to the inactive GDP-bound form (*lower left panel*). The Rho GAP domain is composed almost entirely of α-helices; the position of the catalytic arginine 149 residue is highlighted in *bold*. The C-terminal ADP ribosyltransferase domain (*ADP-R*, residues 236–457) catalyzes the transfer of ADP-ribose from NAD to the mammalian Crk proteins. The catalytic glutamate residues E383 and E385 are highlighted in *bold*. Model of the ExoT Rho GAP domain structure (*lower left panel*, shown in *blue*) was constructed with Swiss PDB viewer by alignment with the ExoS crystal structure. Model of the ExoT ADP ribosyltransferase domain (*lower right panel*, shown in *red*) was constructed with Swiss PDB viewer by alignment of the primary sequences with the known structures of C3, VIP2 and iota toxins

organizes the actin cytoskeleton by inactivating Cdc42, Rac1, and RhoA [30, 33], the significance of the capacity of ExoT to function as a ADP-ribosyltransferase has only recently been recognized. Early studies observed that relative to ExoS, ExoT possessed only limited ADP-ribosyltransferase activity for Ras and SBTI, although ExoS and ExoT shared 76% amino acid identity. Therefore, ExoT was proposed to represent a defective ADP-ribosyltransferase, possibly through gene duplication. Subsequent studies showed that a Rho GAP defective ExoT retained some capacity to reorganize the actin cytoskeleton and contained antiphagocytic activity [21]. Furthermore, ExoT elicited a cell rounding independent of Rho GAP activity and without ADP-ribosylating Ras [68]. This suggested that ExoT ADP-ribosyltransferase activity is active and that ExoT might target host proteins that were distinct from ExoS. Most recently, ExoT was shown to ADP-ribosylate Crk-I and Crk-II both in vitro and in vivo. The rate of ADP-ribosylation of Crk by ExoT is comparable to that of ExoS for SBTI. Therefore, the discovery of ExoT ADP-ribosylation of Crk proteins links the antiphagocytic activity of ExoT to a novel mechanism that is distinct from transient modulation of Rho GTPases by Rho GAP activity. This leads to a hypothesis in which the ADP-ribosylation of Crk proteins by ExoT blocks interactions with either up- or downstream binding partners, thereby blocking Rap1- and Rac1-mediated focal adhesion and phagocytosis signaling.

9
Yersinia Type III Cytotoxins That Modulate the Actin Cytoskeleton: YopE

The initial finding that YopE contributes to the pathogenesis of *Yersinia* was made by Straley and coworkers [66]. Subsequently, Wolf-Watz and coworkers showed that YopE mediated a cytotoxic response on HeLa cells and macrophages.

Binding of *Yersinia* to the host cell surface was required for this intoxication, and YopE appeared to be involved in the inhibition of innate host defenses that lead to an inhibition of phagocytosis [48]. This cytotoxic response was subsequently defined as the reorganization of the actin cytoskeleton and not a direct cytotoxic effect on the cell. Subsequent studies showed that isolated YopE disrupted microfilaments on microinjection, which was an early indication that YopE is directly involved in actin reorganization [49]. These investigators subsequently showed that direct contact between *Yersinia* and the cultured cell induces expression and, subsequently polarized transfer of YopE. In this process, the bacteria remained on the surface of the cultured cell and YopE was transferred across the host cell plasma membrane and could be recovered within the host cell cytosol. This indicated the intricate interactions between the host cell and the type III secretion apparatus [50]. YerA, a chaperone-like protein, was reported to interact specifically with YopE in the bacterium's cytoplasm and contribute to YopE secretion but was dispensable for YopE translocation into cultured cells. This indicated that YerA stabilizes YopE in the bacterium's cytoplasm to maintain YopE in an optimal secretion-competent conformation but that YopE is also capable of chaperone-independent secretion [17].

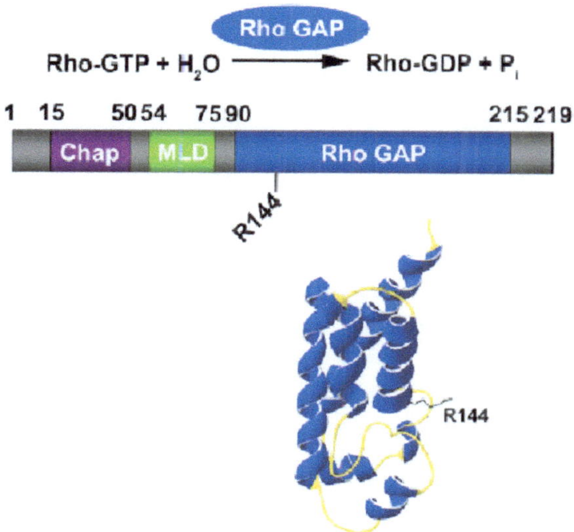

Fig. 3 Structural arrangement of the *Yersinia* effector YopE. The *Yersinia* virulence factor YopE is composed of a series of discreet modules: an N-terminal chaperone binding (*Chap*, residues 15–50) and membrane localization domain (*MLD*, residues 54–75), and a C-terminal Rho GAP domain (residues 90–215). The catalytic arginine 144 residue is highlighted in *bold*. The Rho GAP domain catalyzes the conversion of the active GTP-bound form of Rho, Rac, and Cdc42 to the inactive GDP-bound form. The crystal structure of the GAP domain (*lower panel*, shown in *blue*) shows that the protein is composed almost entirely of α-helices. The position of the catalytic arginine 144 is *highlighted* (YopE from PDB: 1HY5)

YopE is a 219-amino acid protein produced by *Y. pestis*, *Y. pseudotuberculosis*, and *Y. enterocolitica* with essentially identical primary amino acid sequences (Fig. 3). Thus data from YopE isolated from each strain can be combined to provide a detailed description of the structure-function properties of this toxin. Fusion of the *Bordetella pertussis* adenylate cyclase to YopE forms an elegant reporter system to monitor the translocation of YopE into mammalian cells [62]. Using this system, mapping studies demonstrated that the N terminus of YopE was required for both secretion out of *Yersinia* and translocation into the host mammalian cell [56]. The 11 N-terminal residues of YopE were sufficient to secrete the reporter protein out of the bacterium. Although the 49 N-terminal residues were required for translocation into cultured cells, the N-terminal 75 residues included a domain that allowed the release of YopE from the bacterial membrane. These data were an early indication that secretion from the bacterium and translocation into the host mammalian cells are performed by distinct regions within the N terminus of the Yop proteins. Defining the catalytic action of YopE paralleled studies on the *Pseudomonas* ExoS and *Salmonella* SptP type III cytotoxins. YopE was determined to be a Rho GAP for Rho, Rac, and Cdc42 in vitro, which explained the ability of YopE to stimulate actin reorganization in vivo [5]. Rho GAPs enhance the intrinsic GTPase activity of the RhoGTPases by stabilizing a transition state in which the catalytic glutamine residue of the GTPase is correctly aligned to orient the hydrolytic H_2O molecule that extracts the γ-phosphate from GTP, yielding GDP. Like the mammalian Rho GAPs, YopE utilizes an arginine (Arg144) residue to stimulate the hydrolytic activity of the RhoGTPases [5, 7, 72]. The crystal structure of the YopE GAP domain (residues 90–219) displays little homology to its mammalian counterparts apart from the fact that it is composed almost entirely of α-helices. Moreover, the YopE catalytic Arg144 residue is contained on an α-helix, rather than a flexible loop as is the case for all known mammalian GAPs. The physical organization of the YopE Rho GAP domain is essentially identical to that of the ExoS and SptP Rho GAP domains.

10
YopT Is a Cysteine Protease That Cleaves C-Terminal Isoprenoid Group of RhoA

YopT is a 322-amino acid type III cytotoxin of *Yersinia* that was initially shown to disrupt the actin cytoskeleton and cause cell rounding [29] (Fig. 4). Unlike YopE, secretion and translocation of YopT absolutely require the presence of its cognate chaperone, SycT. Disruption of the actin cytoskeleton by YopT contributes to the resistance of *Yersinia* to phagocytosis, because YopT-deficient strains were phagocytosed significantly more efficiently by J774 cells and by human polymorphonuclear leukocytes [1]. The covalent modification of host proteins by bacterial toxins may induce a change in the electrophoretic mobility of the modified host proteins. Similarly, type III delivery of YopT into COS-7 cells resulted in the appearance of a faster-migrating form of RhoA on SDS-PAGE gels and induced an acidic shift in RhoA by isoelectric focusing. Modification of RhoA by YopT facilitated the redistribution of RhoA from the membrane to the cytosol [3]. The altered electrophoretic property of RhoA was an early indication that YopT mediates changes to the actin cytoskeleton through direct covalent modification of the RhoGTPases. Recombinant YopT caused rounding of embryonic bovine lung (EBL) cells and rapid redistribution of the actin cytoskeleton after microinjection. Moreover, incubation of recombinant YopT with EBL cell membranes or recombinant isoprenylated RhoA bound to artificial PE or PE/PIP_2 vesicles caused release

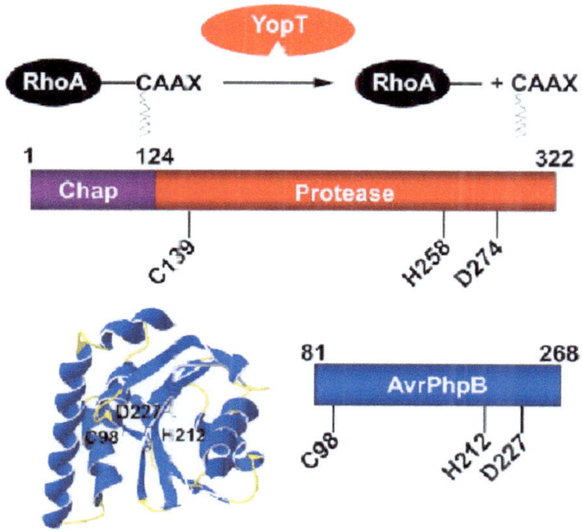

Fig. 4 Structural arrangement of the *Yersinia* effector YopT and the *Pseudomonas syringae* effector AvrPphB. The *Yersinia* virulence factor YopT is composed of a series of discreet modules: an N-terminal chaperone binding domain (*Chap*, residues 1–124) and a C-terminal cysteine protease domain (residues 125–322). The positions of the catalytic triad are highlighted in *bold*. YopT inactivates RhoA through cleavage of the C-terminal CAAX motif, releasing the protein from the membrane. The *Pseudomonas* virulence factor AvrPhpB is a cysteine protease that cleaves the plant kinase PBS1. AvrPhpB belongs to a family of 19 cysteine proteases including YopT. The crystal structure of AvrPhpB (*lower left panel*, shown in *blue*) is composed of central antiparallel β-sheets, with α-helices packing both sides of the β-sheet to form a two-lobe structure. The core of this structure resembles the papainlike cysteine proteases and includes the AvrPphB active site catalytic triad of Cys-98, His-212, and Asp-227 (AvrPphB from PDB: 1UKF)

of the RhoA protein [3]. Dixon and coworkers subsequently demonstrated that YopT cleaves near the carboxyl termini of Rho family GTPases, resulting in the loss of the prenylated cysteine residue from the GTPase [59]. In vitro assays found that YopT cleaves N-terminal to the prenylated cysteine in RhoA, Rac1, and Cdc42 and that the cleavage product of the GTPases is geranylgeranyl cysteine methyl ester. Cleavage of the RhoGTPases was not dependent on the nucleotide state of the GTPase. However, efficient cleavage was dependent on the presence of the lipid modification and required the polybasic region of RhoA for correct recognition [60]. Resent studies by Heesemann and coworkers showed that type III delivered YopT localized to the membranes of mammalian cells and resulted in the release of RhoA, but not Rac1 or Cdc42, from the cell membrane. Moreover, YopT caused the selective disassociation of RhoA from Rho GDI and the accumulation of RhoA in the cytosol. Type III delivery of YopT into macrophages resulted in disruption of actin structures required for the efficient uptake of *Yersinia* [2].

Alignment studies of the primary sequence showed that YopT was related to a family of proteins involved in bacterial pathogenesis, including an avirulence (Avr) protein known as AvrPphB, produced by the plant pathogen *P. syringae*. Type III-delivered AvrPphB is also a cysteine protease that cleaves the host protein kinase PBS1, resulting in cell death [59]. Cleavage of RhoA by YopT and PBS1 by AvrPphB is dependent on the invariant C/H/D residues conserved in the entire YopT family. Xu and coworkers determined the crystal structure of AvrPphB, the YopT homolog [78]. The structure is composed of a central antiparallel β-sheet, with α-helices packing on both sides of the sheet to form a two-lobe structure (Fig. 4). The core of this structure resembles the papainlike cysteine proteases and includes the AvrPphB active site catalytic triad of Cys-98, His-212, and Asp-227. Cornelis and coworkers [57] recently reported that YopE, and to a lesser extent YopT, interferes with the caspase-1 mediated maturation of pro-interleukin-1β. YopE and YopT prevented the autoproteolytic activation of caspase-1 through modulation of Rac1. Rac1 contributed to autoactivation of caspase-1 through activation of the downstream protein LIM kinase-1. LIM kinase-1 is a serine/threonine kinase that phosphorylates the actin regulator cofilin, a major regulator of actin dynamics [77]. Dominant negative forms of Rac1 and LIM kinase-1 inhibit caspase-1 activation in response to both YopE and YopT. These results suggest a new function of Rho GTPases in the regulation of innate immunity and suggest that the *Yersinia* type III cytotoxins, YopE and YopT—perhaps in concert—modulate immune escape by *Yersinia*.

11 YopH Inhibits Phagocytosis Through Dephosphorylating Focal Adhesion Proteins in Crk-Mediated Signal Pathways

YopH is a 468-amino acid type III cytotoxin produced by *Yersinia* that possesses tyrosine phosphatase activity (Fig. 5). It was the first effector shown to possess antiphagocytic effects for *Yersinia*. Contact between the *Yersinia* outer membrane protein invasin and mammalian integrin receptors stimulates rapid tyrosine phosphorylation of several host proteins. Phosphorylation of key cytosolic kinases, such as Src kinase and FAK, promotes the assembly of multimeric focal adhesion complexes containing adaptor molecules such

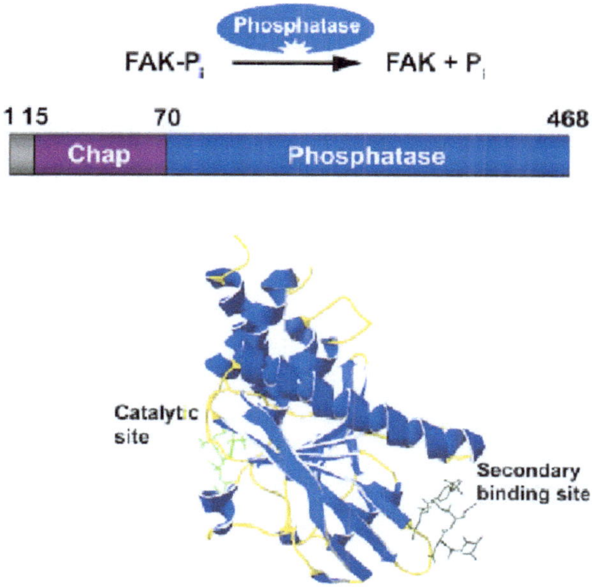

Fig. 5 Structural arrangement of the *Yersinia* effector YopH. The *Yersinia* virulence factor YopH is composed of a series of discreet modules: an N-terminal chaperone binding domain (*Chap*, residues 15–70) and a C-terminal protein phosphatase domain (residues 71–468). YopH disrupts focal adhesion complexes through dephosphorylation of several proteins including focal adhesion kinase (*FAK*). The crystal structure of the YopH phosphatase domain (*lower panel*, shown in *blue*) in complex with an inhibitor (*green*) is shown. The structure identified a primary catalytic site and a secondary, catalytically inactive binding site (YopH from PDB: 1YTS)

as the CT10 regulator of kinase (Crk) proteins. Crk proteins can subsequently recruit the GEF of Rap-1 (CRK SH3 domain-binding guanine nucleotide-releasing factor, C3G) and the GEF of Rac-1 (180-kDa protein downstream of CRK, DOCK180), which activate focal adhesion complexes and phagocytosis, respectively.

Type III-delivered YopH dephosphorylates p130Cas, paxillin, and FAK to inhibit phagocytosis through blockage of the interaction of focal adhesion proteins with Crk proteins. Phan et al. [46] determined the crystal structure of the YopH phosphatase domain complexed with a nonhydrolyzable substrate (Fig. 5). Although ligand binding in the active site occurred by a mechanism similar to the structurally related human phosphatases, YopH also possessed a second peptide-binding site of unknown function. In an independent study, Sun et al. [67] reported the structure of YopH, defining residues involved in substrate interaction and a unique H-bonding network that may contribute to the design of specific YopH inhibitors. Biological studies by Logsdon and Mecsas [37] utilized isogenic strains of *Yersinia* to measure the role of the Yops in colonization and persistence in a mouse model of infection and observed that a *yop*H-mutant strain failed to colonize the mesenteric lymph nodes. This study also reported that although single *yop*-mutant strains did not affect colonization or persistence, a strain that was mutated for both *yop*H and *yop*E failed to colonize intestinal tissues. Thus YopH and YopE may have redundant functions. Recently, Alonso et al. [4] showed that in human T lymphocytes YopH caused a reduction in intracellular tyrosine phosphorylation and inhibited T cell activation. Affinity probes suggested that the Lck tyrosine kinase, the primary signal transducer for the T cell antigen receptor, was dephosphorylated by YopH. Inhibition of Lck by YopH will block early steps in T cell antigen receptor signaling and can inhibit development of a protective immune response.

These findings suggest that the effects of the Yops on host cell physiology extend beyond the primary target of the actin cytoskeleton, and that Yops can work alone or in concert with one another to modulate host physiology. Our understanding of how the Yops contribute to bacterial pathogenesis will parallel advances in the resolution of the hierarchy of the signal transduction cascades, as it appears that Rho GTPases regulate pathways beyond the structural considerations of the actin cytoskeleton.

Acknowledgements The authors are supported by NIH Grants AI-030162 and AI-057153.

References

1. Aepfelbacher M, Trasak C, Wiedemann A, Andor A. 2003. Rho-GTP binding proteins in *Yersinia* target cell interaction. *Adv Exp Med Biol* 529:65–72
2. Aepfelbacher M, Trasak C, Wilharm G, Wiedemann A, Trulzsch K, et al. 2003. Characterization of YopT effects on Rho GTPases in *Yersinia enterocolitica* infected cells. *J Biol Chem* 278:33217–23
3. Aepfelbacher M, Zumbihl R, Ruckdeschel K, Jacobi CA, Barz C, Heesemann J. 1999. The tranquilizing injection of *Yersinia* proteins: a pathogen's strategy to resist host defense. *Biol Chem* 380:795–802
4. Alonso A, Bottini N, Bruckner S, Rahmouni S, Williams S, et al. 2004. Lck dephosphorylation at Tyr-394 and inhibition of T cell antigen receptor signaling by *Yersinia* phosphatase YopH. *J Biol Chem* 279:4922–8
5. Andor A, Trulzsch K, Essler M, Roggenkamp A, Wiedemann A, et al. 2001. YopE of *Yersinia*, a GAP for Rho GTPases, selectively modulates Rac-dependent actin structures in endothelial cells. *Cell Microbiol* 3:301–10.

6. Bergman T, Hakansson S, Forsberg A, Norlander L, Macellaro A, et al. 1991. Analysis of the V antigen lcrGVH-yopBD operon of *Yersinia pseudotuberculosis*: evidence for a regulatory role of LcrH and LcrV. *J Bacteriol* 173:1607–16
7. Black DS, Bliska JB. 2000. The RhoGAP activity of the *Yersinia pseudotuberculosis* cytotoxin YopE is required for antiphagocytic function and virulence. *Mol Microbiol* 37:515–27
8. Castellano F, Chavrier P, Caron E. 2001. Actin dynamics during phagocytosis. *Semin Immunol* 13:347–55
9. Coburn J, Gill DM. 1991. ADP-ribosylation of p21ras and related proteins by *Pseudomonas aeruginosa* exoenzyme S. *Infect Immun* 59:4259–62
10. DeBord KL, Lee VT, Schneewind O. 2001. Roles of LcrG and LcrV during type III targeting of effector Yops by *Yersinia enterocolitica*. *J Bacteriol* 183:4588–98
11. Evdokimov AG, Tropea JE, Routzahn KM, Waugh DS. 2002. Crystal structure of the *Yersinia pestis* GTPase activator YopE. *Protein Sci* 11:401–8
12. Feldman M, Bryan R, Rajan S, Scheffler L, Brunnert S, et al. 1998. Role of flagella in pathogenesis of *Pseudomonas aeruginosa* pulmonary infection. *Infect Immun* 66:43–51
13. Finck-Barbancon V, Goranson J, Zhu L, Sawa T, Wiener-Kronish JP, et al. 1997. ExoU expression by *Pseudomonas aeruginosa* correlates with acute cytotoxicity and epithelial injury. *Mol Microbiol* 25:547–57
14. Finland M. 1972. Changing patterns of susceptibility of common bacterial pathogens to antimicrobial agents. *Ann Intern Med* 76:1009–36
15. Fitzgerald DJ, Fryling CM, Zdanovsky A, Saelinger CB, Kounnas M, et al. 1994. Selection of *Pseudomonas* exotoxin-resistant cells with altered expression of α2MR/LRP. *Ann N Y Acad Sci* 737:138–44
16. Frithz-Lindsten E, Holmstrom A, Jacobsson L, Soltani M, Olsson J, et al. 1998. Functional conservation of the effector protein translocators PopBYopB and PopDYopD of *Pseudomonas aeruginosa* and *Yersinia pseudotuberculosis*. *Mol Microbiol* 29:1155–65
17. Frithz-Lindsten E, Rosqvist R, Johansson L, Forsberg A. 1995. The chaperonelike protein YerA of *Yersinia pseudotuberculosis* stabilizes YopE in the cytoplasm but is dispensable for targeting to the secretion loci. *Mol Microbiol* 16: 635–47
18. Gacesa P. 1998. Bacterial alginate biosynthesis–recent progress and future prospects. *Microbiology* 144:1133–43
19. Ganesan AK, Vincent TS, Olson JC, Barbieri JT. 1999. *Pseudomonas aeruginosa* exoenzyme S disrupts Ras-mediated signal transduction by inhibiting guanine nucleotide exchange factor-catalyzed nucleotide exchange. *J Biol Chem* 274:21823–9
20. Garrity-Ryan L, Kazmierczak B, Kowal R, Comolli J, Hauser A, Engel JN. 2000. The arginine finger domain of ExoT contributes to actin cytoskeleton disruption and inhibition of internalization of *Pseudomonas aeruginosa* by epithelial cells and macrophages. *Infect Immun* 68:7100–13.
21. Garrity-Ryan L, Shafikhani S, Balachandran P, Nguyen L, Oza J, et al. 2004. The ADP ribosyltransferase domain of *Pseudomonas aeruginosa* ExoT contributes to its biological activities. *Infect Immun* 72:546–58
22. Geiser TK, Kazmierczak BI, Garrity-Ryan LK, Matthay MA, Engel JN. 2001. *Pseudomonas aeruginosa* ExoT inhibits in vitro lung epithelial wound repair. *Cell Microbiol* 3:223–36.
23. Goehring UM, Schmidt G, Pederson KJ, Aktories K, Barbieri JT. 1999. The N-terminal domain of *Pseudomonas aeruginosa* exoenzyme S is a GTPase-activating protein for Rho GTPases. *J Biol Chem* 274:36369–72

24. Hakansson S, Schesser K, Persson C, Galyov EE, Rosqvist R, et al. 1996. The YopB protein of *Yersinia pseudotuberculosis* is essential for the translocation of Yop effector proteins across the target cell plasma membrane and displays a contact-dependent membrane disrupting activity. *EMBO J* 15:5812–23
25. Hickling TP, Sim RB, Malhotra R. 1998. Induction of TNF-α release from human buffy coat cells by *Pseudomonas aeruginosa* is reduced by lung surfactant protein A. *FEBS Lett* 437:65–9
26. Hueck CJ. 1998. Type III protein secretion systems in bacterial pathogens of animals and plants. *Microbiol Mol Biol Rev* 62:379–433
27. Iglewski BH, Liu PV, Kabat D. 1977. Mechanism of action of *Pseudomonas aeruginosa* exotoxin Ai adenosine diphosphate-ribosylation of mammalian elongation factor 2 in vitro and in vivo. *Infect Immun* 15:138–44
28. Iglewski BH, Sadoff J, Bjorn MJ, Maxwell ES. 1978. *Pseudomonas aeruginosa* exoenzyme S: an adenosine diphosphate ribosyltransferase distinct from toxin A. *Proc Natl Acad Sci USA* 75:3211–5
29. Iriarte M, Cornelis GR. 1998. YopT, a new *Yersinia* Yop effector protein, affects the cytoskeleton of host cells. *Mol Microbiol* 29:915–29
30. Kazmierczak BI, Engel JN. 2002. *Pseudomonas aeruginosa* ExoT acts in vivo as a GTPase-activating protein for RhoA, Rac1, and Cdc42. *Infect Immun* 70:2198–205
31. Knight DA, Finck-Barbancon V, Kulich SM, Barbieri JT. 1995. Functional domains of *Pseudomonas aeruginosa* exoenzyme S. *Infect Immun* 63:3182–6
32. Koster M, Bitter W, de Cock H, Allaoui A, Cornelis GR, Tommassen J. 1997. The outer membrane component, YscC, of the Yop secretion machinery of *Yersinia enterocolitica* forms a ring-shaped multimeric complex. *Mol Microbiol* 26:789–97
33. Krall R, Schmidt G, Aktories K, Barbieri JT. 2000. *Pseudomonas aeruginosa* ExoT is a Rho GTPase-activating protein. *Infect Immun* 68:6066–8.
34. Krall R, Sun J, Pederson KJ, Barbieri JT. 2002. In vivo rho GTPase-activating protein activity of *Pseudomonas aeruginosa* cytotoxin ExoS. *Infect Immun* 70: 360–7
35. Krall R, Zhang Y, Barbieri JT. 2004. Intracellular membrane localization of *Pseudomonas* ExoS and *Yersinia* YopE in mammalian cells. *J Biol Chem* 279: 2747–53
36. Lee VT, Tam C, Schneewind O. 2000. LcrV, a substrate for *Yersinia enterocolitica* type III secretion, is required for toxin targeting into the cytosol of HeLa cells. *J Biol Chem* 275:36869–75
37. Logsdon LK, Mecsas J. 2003. Requirement of the *Yersinia pseudotuberculosis* effectors YopH and YopE in colonization and persistence in intestinal and lymph tissues. *Infect Immun* 71:4595–607
38. McIver KS, Kessler E, Olson JC, Ohman DE. 1995. The elastase propeptide functions as an intramolecular chaperone required for elastase activity and secretion in *Pseudomonas aeruginosa*. *Mol Microbiol* 18:877–89
39. Mendelson MH, Gurtman A, Szabo S, Neibart E, Meyers BR, et al. 1994. *Pseudomonas aeruginosa* bacteremia in patients with AIDS. *Clin Infect Dis* 18: 886–95
40. Meyer JM, Stintzi A, Poole K. 1999. The ferripyoverdine receptor FpvA of *Pseudomonas aeruginosa*PAO1 recognizes the ferripyoverdines of *P. aeruginosa* PAO1 and *P. fluorescens* ATCC 13525. *FEMS Microbiol Lett* 170:145–50
41. Michiels T, Wattiau P, Brasseur R, Ruysschaert JM, Cornelis G. 1990. Secretion of Yop proteins by *Yersiniae*. *Infect Immun* 58:2840–9
42. Neyt C, Cornelis GR. 1999. Insertion of a Yop translocation pore into the macrophage plasma membrane by *Yersinia enterocolitica*: requirement for translocators YopB and YopD, but not LcrG. *Mol Microbiol* 33:971–81

43. Nobes CD, Hall A. 1995. Rho, rac, and cdc42 GTPases regulate the assembly of multimolecular focal complexes associated with actin stress fibers, lamellipodia, and filopodia. *Cell* 81:53–62
44. Pederson KJ, Krall R, Riese MJ, Barbieri JT. 2002. Intracellular localization modulates targeting of ExoS, a type III cytotoxin, to eukaryotic signalling proteins. *Mol Microbiol* 46:1381–90
45. Pederson KJ, Vallis AJ, Aktories K, Frank DW, Barbieri JT. 1999. The amino terminal domain of *Pseudomonas aeruginosa* ExoS disrupts actin filaments via small-molecular-weight GTP-binding proteins. *Mol Microbiol* 32:393–401
46. Phan J, Lee K, Cherry S, Tropea JE, Burke TR, Jr., Waugh DS. 2003. High resolution structure of the *Yersinia pestis* protein tyrosine phosphatase YopH in complex with a phosphotyrosyl mimetic-containing hexapeptide. *Biochemistry* 42:13113–21
47. Ridley AJ. 2001. Rho family proteins: coordinating cell responses. *Trends Cell Biol* 11:471–7
48. Rosqvist R, Forsberg A, Rimpilainen M, Bergman T, Wolf-Watz H. 1990. The cytotoxic protein YopE of *Yersinia* obstructs the primary host defence. *Mol Microbiol* 4:657–67
49. Rosqvist R, Forsberg A, Wolf-Watz H. 1991. Intracellular targeting of the *Yersinia* YopE cytotoxin in mammalian cells induces actin microfilament disruption. *Infect Immun* 59:4562–9
50. Rosqvist R, Magnusson KE, Wolf-Watz H. 1994. Target cell contact triggers expression and polarized transfer of *Yersinia* YopE cytotoxin into mammalian cells. *EMBO J* 13:964–72
51. Rust L, Pesci EC, Iglewski BH. 1996. Analysis of the *Pseudomonas aeruginosa* elastase (lasB) regulatory region. *J Bacteriol* 178:1134–40
52. Sakagawa E. 1998. [Study on the role of mucoid strain in chronic airway infections]. *Kansenshogaku Zasshi* 72:379–94
53. Sarker MR, Neyt C, Stainier I, Cornelis GR. 1998. The *Yersinia* Yop virulon: LcrV is required for extrusion of the translocators YopB and YopD. *J Bacteriol* 180:1207–14
54. Sathish K, Padma B, Munugalavadla V, Bhargavi V, Radhika KV, et al. 2004. Phosphorylation of profilin regulates its interaction with actin and poly(L proline). *Cell Signal* 16:589–96
55. Sawa T, Yahr TL, Ohara M, Kurahashi K, Gropper MA, et al. 1999. Active and passive immunization with the *Pseudomonas* V antigen protects against type III intoxication and lung injury [see comments]. *Nat Med* 5:392–8
56. Schesser K, Frithz-Lindsten E, Wolf-Watz H. 1996. Delineation and mutational analysis of the *Yersinia pseudotuberculosis* YopE domains which mediate translocation across bacterial and eukaryotic cellular membranes. *J Bacteriol* 178: 7227–33
57. Schotte P, Denecker G, Van Den Broeke A, Vandenabeele P, Cornelis GR, Beyaert R. 2004. Targeting Rac1 by the *Yersinia* effector protein YopE inhibits caspase-1 mediated maturation and release of interleukin-1β (IL-1β). *J Biol Chem*
58. Schweizer F, Jiao H, Hindsgaul O, Wong WY, Irvin RT. 1998. Interaction between the pili of *Pseudomonas aeruginosa* PAK and its carbohydrate receptor β-D-GalNAc(1$\rightarrow$4)β-D-Gal analogs. *Can J Microbiol* 44:307–11
59. Shao F, Merritt PM, Bao Z, Innes RW, Dixon JE. 2002. A *Yersinia* effector and a *Pseudomonas* avirulence protein define a family of cysteine proteases functioning in bacterial pathogenesis. *Cell* 109:575–88
60. Shao F, Vacratsis PO, Bao Z, Bowers KE, Fierke CA, Dixon JE. 2003. Biochemical characterization of the *Yersinia* YopT protease: cleavage site and recognition elements in Rho GTPases. *Proc Natl Acad Sci USA* 100:904–9

61. Skrzypek E, Straley SC. 1995. Differential effects of deletions in lcrV on secretion of V antigen, regulation of the low-Ca^{2+} response, and virulence of *Yersinia pestis*. *J Bacteriol* 177:2530–42
62. Sory MP, Cornelis GR. 1994. Translocation of a hybrid YopE-adenylate cyclase from *Yersinia enterocolitica* into HeLa cells. *Mol Microbiol* 14:583–94
63. Stanislavsky ES, Lam JS. 1997. *Pseudomonas aeruginosa* antigens as potential vaccines. *FEMS Microbiol Rev* 21:243–77
64. Stebbins CE, Galan JE. 2000. Modulation of host signaling by a bacterial mimic: structure of the *Salmonella* effector SptP bound to Rac1. *Mol Cell* 6:1449–60
65. Stover CK, Pham XQ, Erwin AL, Mizoguchi SD, Warrener P, et al. 2000. Complete genome sequence of *Pseudomonas aeruginosa* PA01, an opportunistic pathogen. *Nature* 406:959–64
66. Straley SC, Cibull ML. 1989. Differential clearance and host-pathogen interactions of YopE- and YopK-YopL- *Yersinia pestis* in BALB/c mice. *Infect Immun* 57:1200–10
67. Sun JP, Wu L, Fedorov AA, Almo SC, Zhang ZY. 2003. Crystal structure of the *Yersinia* protein-tyrosine phosphatase YopH complexed with a specific small molecule inhibitor. *J Biol Chem* 278:33392–9
68. Sundin C, Henriksson ML, Hallberg B, Forsberg A, Frithz-Lindsten E. 2001. Exoenzyme T of *Pseudomonas aeruginosa* elicits cytotoxicity without interfering with Ras signal transduction. *Cell Microbiol* 3:237–46.
69. Takeda H. 1998. [Study on pathogenetic role of myeloperoxidase, tumor necrosis factor α, interferon γin chronic airway infection with *Pseudomonas aeruginosa*]. *Kansenshogaku Zasshi* 72:395–409
70. Vasil ML, Graham LM, Ostroff RM, Shortridge VD, Vasil AI. 1991. Phospholipase C: molecular biology and contribution to the pathogenesis of *Pseudomonas aeruginosa*. *Antibiot Chemother* 44:34–47
71. Vasil ML, Ochsner UA, Johnson Z, Colmer JA, Hamood AN. 1998. The furregulated gene encoding the alternative sigma factor PvdS is required for iron-dependent expression of the LysR-type regulator ptxR in *Pseudomonas aeruginosa*. *J Bacteriol* 180:6784–8
72. Von Pawel-Rammingen U, Telepnev MV, Schmidt G, Aktories K, Wolf-Watz H, Rosqvist R. 2000. GAP activity of the *Yersinia* YopE cytotoxin specifically targets the Rho pathway: a mechanism for disruption of actin microfilament structure. *Mol Microbiol* 36:737–48
73. Woestyn S, Allaoui A, Wattiau P, Cornelis GR. 1994. YscN, the putative energizer of the *Yersinia* Yop secretion machinery. *J Bacteriol* 176:1561–9
74. Wurtele M, Wolf E, Pederson KJ, Buchwald G, Ahmadian MR, et al. 2001. How the *Pseudomonas aeruginosa* ExoS toxin downregulates Rac. *Nat Struct Biol* 8:23–6.
75. Yahr TL, Mende-Mueller LM, Friese MB, Frank DW. 1997. Identification of type III secreted products of the *Pseudomonas aeruginosa* exoenzyme S regulon. *J Bacteriol* 179:7165–8
76. Yahr TL, Vallis AJ, Hancock MK, Barbieri JT, Frank DW. 1998. ExoY, an adenylate cyclase secreted by the *Pseudomonas aeruginosa* type III system. *Proc Natl Acad Sci USA* 95:13899–904
77. Yang N, Higuchi O, Ohashi K, Nagata K, Wada A, et al. 1998. Cofilin phosphorylation by LIM-kinase 1 and its role in Rac-mediated actin reorganization. *Nature* 393:809–12
78. Zhu M, Shao F, Innes RW, Dixon JE, Xu Z. 2004. The crystal structure of *Pseudomonas* avirulence protein AvrPphB: a papain-like fold with a distinct substrate-binding site. *Proc Natl Acad Sci USA* 101:302–26

CTMI (2005) 291:167–175

Modulation of Rho GTPases and the Actin Cytoskeleton by YopT of *Yersinia*

M. Aepfelbacher[1,3] · R. Zumbihl[2] · J. Heesemann[1] (✉)

[1]Lehrstuhl Bakteriologie, Max von Pettenkofer-Institut für Hygiene und Medizinische Mikrobiologie, Pettenkoferstr. 9a, 80336 München, Germany
heesemann@m3401.mpk.med.uni-muenchen.de

[2]Laboratoire Ecologie microbienne des Insectes et Interaction Hôte-Pathogène (EMIP), Université Montpellier II, C.C. 101, Bat 24, Unité INRA n°1133, 34095 Montpellier Cedex 05, France

[3]*Present Address:* Institut für Infektionsmedizin, Universitätsklinikum Hamburg-Eppendorf, Martinistr. 52, 20246 Hamburg, Germany

Abstract Pathogenic *Yersinia* species evade the innate cellular immune response by injecting antihost effector proteins (*Yersinia* outer proteins, Yops) into host cells through a type III secretion (TTS) apparatus. One of the six effector Yops, YopT, inactivates the small GTPase RhoA by removing the geranylgeranylated C-terminal cysteine. This cleavage results in release of RhoA from the cell membrane and subsequently in blockage of stress fiber formation. Thus YopT impairs cellular functions associated with cytoskeleton rearrangements.

1 Introduction

The physiological bacterial flora and the closely related pathogens of a host organism share common aims: survival, persistence and replication in the host, and transmission to new hosts. However, the pathogens are more reckless; they break through the physical barriers of the host and occupy niches that are too hostile for nonpathogenic bacteria (Heesemann 2002). Although the host senses pathogenic as well as nonpathogenic invaders, it is the pathogen that resists or evades the innate immune defense. Obviously, pathogenic microorganisms have developed highly sophisticated strategies to protect themselves against humoral antimicrobial factors of the host and to take over the control of cellular host defense (reviewed by Knodler et al. 2001). To accomplish this, pathogenic microorganisms have acquired sets of large gene blocks (pathogenicity islands and virulence plasmids) by horizontal transfer from distinct unrelated microorganisms, which provide them with divers antihost weapons (Hacker and Kaper 2000). This is also true for the three pathogenic species of the *genus Yersinia*: the bubonic plague bacillus *Y. pestis*, and the enteropathogenic *Y. pseudotuberculosis* and *Y. enterocolitica*. These pathogenes have acquired closely related 70-kb virulence plasmids (called pYV), which provides them with the capability to invade, survive, and replicate extracellularly in host tissue (in particular lymphatic tissue). The virulence plasmid pYV encodes for structural proteins that form a protein microinjection apparatus, the so-called type III secretion (TTS) machine of *Yersinia* (for review see Cornelis 2001). Moreover, pYV carries genes for about 14 known TTS substrates (secreted proteins), of which 7 represent antihost effector proteins (the V-antigen LcrV and the *Yersinia* outer proteins YopE, YopH, YopM, YopO/YpkA, YopP/YopJ, and YopT), whereas the remaining secreted proteins are involved in regulation, protein secretion, and translocation of Yops across the cytoplasmic membrane of host cells. Microinjection of Yops is triggered by intimate contact of *Yersiniae* to host cells (Fig. 1, Rosqvist et al. 1994). As Yops are stored in the cytosol of the pathogen as secretion-competent Yop-chaperone complexes (specific Yop chaperone, Syc), the initial translocation of Yops into host cells might be a burstlike microinjection. This appears to be reasonable when phagocytic processes and the oxidative burst of professional phagocytes must be inhibited rapidly. Moreover, the six translocated Yops target different signaling and antimicrobial activation pathways accumulating in a synergistic antihost effect (reviewed by Aepfelbacher and Heesemann 2002; Bliska 2000; Cornelis 2002; Cornelis and Wolf-Watz 1997; Juris and Dixon 2002). YopH acts as protein tyrosine phosphatase for focal adhesion kinase (FAK) and $p130^{Cas}$, the Src kinase family member Lck-1 and several regulator proteins of the focal adhesion complexes and T-/B-cell receptor signaling complexes (Alonso et al. 2004; Yao et al. 1999). YopP/YopJ inhibits the activation of mitogen- and stress-activated kinases such as ERK 1/2, JNK, and p38 and the nuclear transcription factor NF-κB. The detailed function of YopM remains elusive. YopE, YopT, and YopO/YpkA target small GTPases of Rho family members (RhoA and Rac). YopE's amino acid sequence revealed a GTPase-activating motif, which explains how YopE may shut off Rac activity. YopO (*Y. enterocolitca*) and YpkA (YopO homolog in *Y. pseudotuberculosis*), respectively, is a serine/threonine kinase that is ac-

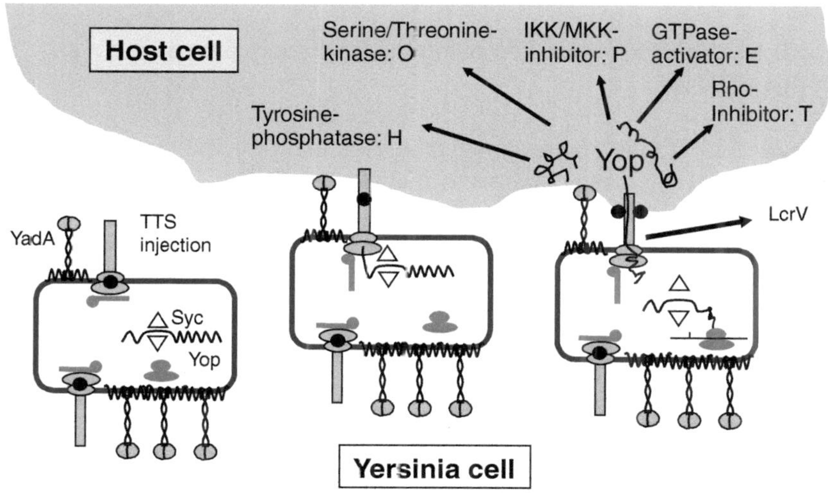

Fig. 1 Schematic drawing of contact-induced Yop secretion/translocation by the *Yersinia* type III-secretion machine. Yops are stabilized by bound chaperones (*Syc*) in the cytosol. *Yersinia*-host cell contact is mediated by the adhesin YadA (a lollipop-shaped proteinaceous projection anchored in the outer membrane and covering entirely the surface of the microbe). *TTS*, type III secretion system; *MKK*, MAP kinase kinase; *IKKβ*, inhibitory κB kinase *β*; *LcrV*, V-antigen; for further information, see text

tivated by monomeric actin and phosphorylates itself and artificial basic substrates. No physiological substrates of YopO have been identified yet. However, YopO/YpkA interacts with RhoA and Rac independent of their GDP/GTP binding state.

This paper will deal with the recent advances of the cell biological effects, biochemical function, and structural features of YopT. Originally, YopT was identified as the second Yop besides YopE that disrupts the actin cytoskeleton, suggesting redundancy of Yop effectors. This suggestion was supported by the fact that more than 50% of *Y. pseudotuberculosis* strains do not carry a functional *yopT* gene and that deletion of the *yopT* gene of *Y. enterocolitica* does not result in virulence attenuation (Iriarte and Cornelis 1998; Trülzsch et al. 2004). However, it was shown later that YopE and YopT inactivate RhoGTPases by different mechanisms. In contrast to YopE, which acts a GTPase activating protein for Rho GTPases, the latter are irreversibly inactivated by YopT, which proteolytically removes the isoprenylated C-terminal cysteine residue.

2 YopT: A Cysteine Protease Removing the Isoprenoid Group of Rho GTPases

Yersiniae inject only minute amounts of Yops into target cells, and intracellular Yop activity is temporally and spatially controlled. Therefore, cellular infection models provide specific informations as to the physiological substrates and biochemical functions of Yops. In fact, destruction of actin stress fibers in *Y. enterocolitica*-infected cells was the first activity pointing to the function of YopT (Iriarte and Cornelis 1998). The basis for this YopT effect was subsequently shown to be modification and inactivation of the Rho GTPase RhoA (Zumbihl et al. 1999). Thereafter the biochemical activity of YopT as protease removing the C-terminal isoprenoid group of RhoGTPases was discovered (Shao et al. 2002, 2003).

YopT is a 35-kDa protein (322 amino acids) that belongs to the CA clan of cysteine proteases. It is expressed by *Y. enterocolitica*, *Y. pestis*, and only some *Y. pseudotuberculosis* strains (Iriarte and Cornelis 1998; Shao et al. 2002; Zumbihl et al. 1999). In addition to three *Yersinia yopT* sequences, 16 homologous open reading frames derived from animal pathogens such as *Haemophilus ducreyi*, *E. coli* O157:H7, or *Pasteurella multocida*, plant pathogens such as *Pseudomonas syringae pv. phaseolicola*, and entomopathogenic *Photorhabdus luminescens* have been extracted from databases. In contrast to the *Photorhabdus luminescens* gene, which is predicted to encode a protein highly similar to YopT (Brugirard-Ricaud et al. 2004), the other sequences are highly diverse. The common feature of all YopT family members is the presence of conserved C/H/D amino acid residues that appear to be essential for protein activity. The cDNA-inferred YopT family proteins can be divided into two groups. The first group that includes YopT contains proteins of 30–40 kDa; the second group contains proteins of >300 kDa that harbor additional functional domains (Shao et al. 2002).

Cellular overexpression and in vitro studies showed that YopT proteolytically removes the geranylgeranyl isoprenoid moiety of RhoA, Rac1, and Cdc42, and this activity is dependent on the invariant C/H/D residues C139, H258, and D274. YopT cleaves just before the C-terminal cysteine to which the geranylgeranyl group is attached via a thioether bond (Shao et al. 2002, 2003). Catalytically inactive YopTC139S can still bind to RhoA and thus can be used for pull-down experiments (Aepfelbacher et al. 2003; Shao et al. 2002; Sorg et al. 2001). Although YopT requires an isoprenoid group for binding and activity, it does not distinguish between geranylgeranylated or farnesylated RhoA. It also seems to work equally well on GDP- and GTP-bound RhoA.

In the test tube and in cells overexpressing the reaction components YopT works best on RhoA, is less active on Rac and CDC42, and does not cleave H-Ras. When the five basic amino acid residues (3 x lysine and 2 x arginine) at the C-terminus of RhoA were mutated to glutamine residues, the resulting RhoA mutant became insensitive to YopT cleavage. Furthermore, an intracellularly expressed GFP fusion construct containing the last 13 amino acids of RhoA (including the 5 basic residues) but not a construct containing the last 4 amino acids of RhoA was cleaved by YopT, although both GFP constructs became isoprenylated (Shao et al. 2002, 2003b). Together these data suggest that YopT recognizes an isoprenoid group in combination with a stretch of basic amino acid at the C-terminus of Rho GTPases. The structural requirements of YopT for binding and cleavage of RhoA were also tested. Whereas deletion of 8 amino acids from the C-terminus abrogated YopT activity, deletion of 74 amino acids from the N-terminus had no effect. When more than 100 amino acids were deleted from the N-terminus of YopT, its activity was abolished. In comparison, binding to RhoA was not greatly affected by deleting 14 amino acids from the C-terminus, whereas the N-terminal 74-amino acid deletion mutant displayed considerably reduced RhoA binding. These findings suggest that catalytic activity is mainly located at the C-terminus, whereas substrate binding also involves the very N-terminus of YopT (Sorg et al. 2003).

3
Cellular Effects of Translocated YopT

RhoA modification by removal of the isoprenoid group has a variety of consequences in *Yersinia*-infected cells (Fig. 2). RhoA is released from the plasma membrane and from its cytoplasmic binding partner guanine nucleotide dissociation inhibitor-1 (GDI-1) and accumulates as a monomeric protein in the cytoplasm. Notably, neither Rac1 nor Cdc42 is removed from cell membranes or GDI-1, suggesting that YopT does not work on these proteins in infected cells (Aepfelbacher et al. 2003; Zumbihl et al. 1999). As part of a systematic approach, the role of YopT in preventing opsonized and unopsonized phagocytosis of *Yersinia enterocolitica* by human neutrophils and mouse macrophages was also investigated. Mutant bacteria lacking YopT were phagocytosed significantly more than wild-type bacteria under both opsonizing and nonopsonizing conditions. *Yersinia* mutants translocating only YopT were not resistant to phagocytosis by neutrophils or macrophages (Grosdent et al. 2002). However, in primary macrophages YopT-overexpressing mutants disrupted actin-rich phagocytic cups induced by *Yersinia* invasin as well as podosomal adhesion structures required for chemotaxis (Aepfelbacher et al.

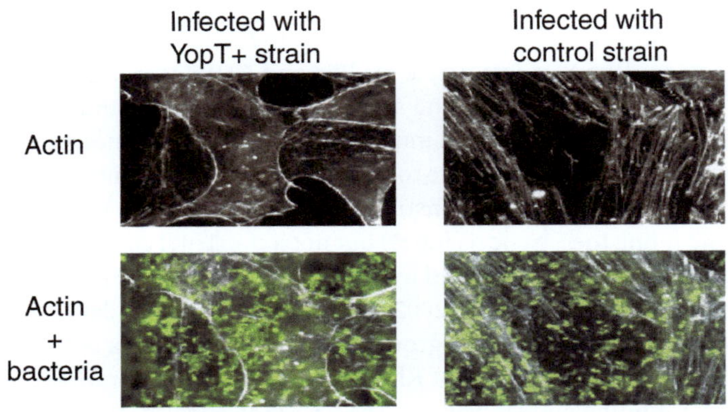

Fig. 2 Effect of YopT on actin stress fibers. Endothelial cells (*HUVEC*) were infected with a control strain [WAC(pYLCR)], lacking *yopT* gene or a strain injecting YopT [WAC(pYLCR+T)]. Cells were then stimulated with thrombin to form actin stress fibers. Filamentous actin was stained with rhodamine phalloidin and bacteria with an anti-*Yersinia* antibody followed by a secondary FITC-conjugated antibody. Stress fibers were readily formed in cells infected with the control strain, whereas their formation was blocked in cells infected with the YopT-expressing strain. (For details see Aepfelbacher et al. 2003)

2003). Hence, YopT alone and in combination with partner Yops can disrupt immune cell function, thereby promoting *Yersinia* infection. At least in some cell types RhoA seems to be the major target of YopT, whereas Rac and CDC42 are not affected. However, at present it cannot be excluded that YopT has additional substrates that may be within the large RhoGTPase family or even unrelated to it.

In infected cells, YopT is located at membranes whereas the majority of its major target RhoA is complexed to GDI in the cytosol. Furthermore in vitro YopT can modify RhoA complexed to GDI-1 only when additional factors such as the membrane lipid phosphatidylinositol bisphosphate (PIP_2) are present or RhoA is artificially loaded with GTP-γS (Aepfelbacher et al. 2003). Thus YopT likely requires additional signaling molecules (such as PIP_2 production and GEFs) to modify RhoA in cells. Taking all these data together, a model of YopT function within *Yersinia*-infected cells is proposed that includes temporal and spatial considerations: YopT translocated into target cells by the *Yersinia* TTS binds to the plasma membrane, where it associates with RhoA and cleaves off its isoprenoid membrane anchor. The truncated RhoA is trapped in the cytosol. By a mechanism likely involving membrane lipids and exchange factors RhoA is released from GDI and then translocates directly to YopT or to membranes where it is cleaved by YopT. This cycle proceeds until all of the RhoA is modified (Fig. 3).

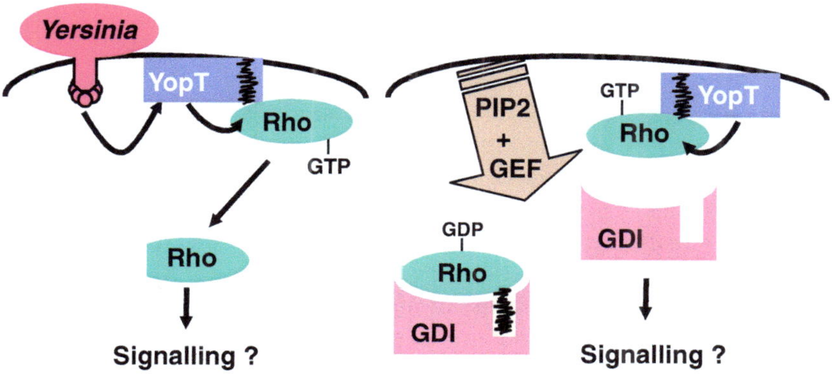

Fig. 3 YopT function inside cells. The effects of YopT in cells infected with *Yersiniae* is depicted taking into account temporal and spatial considerations. *Left*: On injection by the *Yersinia* TTS YopT locates to the plasma membrane, where it binds RhoA via its isoprenoid membrane anchor. Through proteolytic removal of the isoprenoid group RhoA is released into the cytosol. *Right*: Through action of PIP_2 and/or guanine nucleotide exchange factors (GEF's) RhoA is released from its cytosolic complex with GDI and then cleaved by YopT. (For details see text Aepfelbacher et al. 2003)

4
AvrPpkB, the Phytopathogen Homolog of YopT

A different mode of action has been revealed for the YopT-like AvrPphB protein from the plant pathogen *Pseudomonas syringae*. AvrPphB belongs to the group of avirulence proteins that induce a hypersensitivity reaction (HR) at the infection site in resistant plants. HR culminates in localized programmed cell death, thus limiting further pathogen growth and spread. Plant R-genes are known to mediate the HR by sensing the avirulence proteins, which appear to be translocated by the bacterial TTSS. AvrPphB has a proteolytic activity that is required for both autoproteolysis and cleavage of PBS1, a plant protein kinase. Autoproteolysis of the 35-kDa AvrPphB protein between lysine-62 and glycine-63 uncovers a myristoylation motif in an active 28-kDa AvrPphB fragment (Shao et al. 2003). Cleavage of PBS1, on the other hand, is a prerequisite for detection of the AvrPphB protein by the product of the plant R-gene RPS5. Thus the AvrPphB protein is recognized in plants as a consequence of its virulence activity, i.e., proteolytic cleavage of its substrate PBS1.

5 Conclusions

Within the last couple of years much has been learned about the biochemistry and cell biology of YopT and its homolog AvrphB. An exact structural understanding of how YopT cleaves Rho GTPases will require crystallographic data. The crystal structure of the remote YopT homolog AvrPphB has been solved and reveals similarities with papainlike cysteine proteases. A model for the substrate binding mechanism of AvrPphB was suggested. Consistent with the differing substrate specificities of YopT and AvrPphB, the residues corresponding to the substrate binding sites are highly divergent among the YopT family proteins (Zhu et al. 2004), indicating a modular structure typical for pathogenicity factors that have a long history of host-pathogen coevolution.

Unraveling the molecular mechanism of Yop effector function may be helpful to elucidate not only the pathomechanism of *Yersinia* infection but also the infection strategy of phyto- and entomopathogens. Finally, *Yersinia* Yops, in particular those modulating Rho family members, may provide cell biologists with new tools to study the dynamics of cytoskeleton rearrangements and signal transduction processes.

Acknowledgements Work of M.A. and J.H. cited herein was supported by grants from the Deutsche Forschungsgemeinschaft. Work of R.Z. was supported by the Institute National de la Recherche Agronomique and the Ministère de l'Industrie et de Finances (Après séquencage des Génomes).

References

Aepfelbacher M, Heesemann J (2001) Modulation of Rho GTPases and the actin cytoskeleton by *Yersinia* outer proteins (Yops). Int J Med Microbiol 291:269–276

Aepfelbacher M, Trasak C, Wilharm G, Wiedemann A, Trulzsch K, Krauss K, Gierschik P, Heesemann J (2003) Characterization of YopT effects on Rho GTPases in *Yersinia enterocolitica*-infected cells. J Biol Chem: 278:33217–23

Aktories K, Schmidt G, Just I. (2000) Rho GTPases as targets of bacterial protein toxins. Biol Chem 381:421–426

Alonso A, Bottini N, Bruckner S, Rahmouni S, Williams S, Schoenberger SP, Mustelin R (2004) Lck dephosphorylation at Tyr-394 and inhibition of T cell antigen receptor signaling by *Yersinia* phosphatase YopH. J Biol Chem 279:4922–4928

Bishop AL, Hall A (2000) Rho GTPases and their effector proteins. Biochem J 348:241–255

Bliska JB (2000) Yop effectors of *Yersinia* spp. and actin rearrangements. Trends Microbiol 8:205–208

Brugirard-Ricaud K, Givaudan A, Parkhill J, Boemare N, Kunst F, Zumbihl R, Duchaud E (2004) Variation in the effectors of the type III secretion system among *Photorabdus* species as revealed by genomic analysis. J Bacteriol 186:4376–4381

Burridge K, Wennerberg K (2004) Rho and Rac take center stage. Cell 116:167–179
Cornelis GR (2002) The *Yersinia* Ysc-Yop 'T' Weaponry. Nat Rev Mol Cell Biol: 3:742–752
Cornelis GR, Wolf-Watz H (1997) The *Yersinia* Yop virulon: a bacterial system for subverting eukaryotic cells. Mol Microbiol 23:861–867
Grosdent N, Maridonneau-Parini I, Sory MP, Cornelis GR (2002) Role of Yops and adhesins in resistance of *Yersinia enterocolitica* to phagocytosis. Infect Immun 70:4165–76
Hacker J, Kaper JB (2000) Pathogenicity island and the evolution of microbes. Annu Rev Microbiol 54:641–679
Heesemann J (2002) Host defenses against microorganisms: nonspecific defenses. In: Hacker J, Heesemann J (eds) Molecular Infection Biology. Wiley-Liss/Spektrum-Verlag, Heidelberg, pp 29–43
Iriarte M, Cornelis GR (1998) YopT, a new *Yersinia* effector protein, affects the cytoskeleton of host cells. Mol Microbiol 29:915–929
Juris SJ, Shao F, Dixon FE (2002) *Yersinia* effectors target mammalian signalling pathways. Cell Microbiol 4:201–211
Knodler LA, Celli J, Finlay BB (2001) Pathogenic trickery: deception of host cell processes. Nature Rev Mol Cell Biol 2:578–588
Persson C, Nordfelth R, Andersson K, Forsberg A, Wolf-Watz H, Fallman M (1999) Localization of the *Yersinia* PTPase to focal complexes is an important virulence mechanism. Mol Microbiol 33:828–838
Rosqvist R, Magnusson KE, Wolf-Watz H (1994) Target cell contact triggers expression and polarized transfer of *Yersinia* YopE cytotoxin into mammalian cells. EMBO J 13:964–972
Shao F, Golstein C, Ade J, Stoutemyer M, Dixon JE, Innes RW (2003) Cleavage of *Arabidopsis* PBS1 by a bacterial type III effector. Science 301:1230–1233
Shao F, Merritt PM, Bao Z, Innes RW, Dixon JE (2002) A *Yersinia* effector and a *Pseudomonas* avirulence protein define a family of cysteine proteases functioning in bacterial pathogenesis. Cell 109:575–588
Shao F, Vacratsis PO, Bao Z, Bowers KE, Fierke CA, Dixon JE (2003) Biochemical characterization of the *Yersinia* YopT protease: cleavage site and recognition elements in Rho GTPases. Proc Natl Acad Sci USA 100:904–909
Sorg I, Hoffmann C, Dumbach J, Aktories K, Schmidt G (2003) The C terminus of YopT is crucial for activity and the N terminus is crucial for substrate binding. Infect Immun 71:4623–32
Symons M, Settleman J (2000) Rho family GTPases: more than simple switches. Trends Cell Biol 10:415–419
Trülzsch K, Sporleder T, Igwe E, Rüssmann H, Heesemann J (2004) Contribution of the major secreted effector proteins (Yops) of *Yersinia enterocolitica* O:8 to pathogenicity in the mouse infection model. Infect. Immun 72:5227–5234
Van Aelst L, D'Souza-Schorey C (1997) Rho GTPases and signaling networks. Genes Dev. 11:2295–2322
Yao T, Mecsas J, Healy JI, Falkow S, Chien Y (1999) Suppression of T and B lymphocyte activation by a *Yersinia pseudotuberculosis* virulence factor, YopH. J Exp Med. 190:1343–1350
Zhu M, Shao F, Innes RW, Dixon JE, Xu Z (2004) The crystal structure of *Pseudomonas* avirulence protein AvrPphB: a papain-like fold with a distinct substrate-binding site. Proc Natl Acad Sci USA 101:302–307
Zumbihl R, Aepfelbacher M, Andor A, Jacobi CA, Ruckdeschel K, Rouot B, Heesemann J (1999) The cytotoxin YopT of *Yersinia enterocolitica* induces modification and cellular redistribution of the small GTP-binding protein RhoA. J Biol Chem 274:29289–29293

CTMI (2005) 291:177–190

Bacterial Toxins Activating Rho GTPases

P. Munro · E. Lemichez (✉)

Faculté de Médecine, 1/ INSERM, U627, 23 Avenue de Valombrose, 06107 Nice Cedex 2, France
lemichez@unice.fr

Abstract The CNF1 toxin is produced by some uropathogenic (UPECs) and meningitis-causing *Escherichia coli* strains. It belongs to a large family of bacterial virulence factors and toxins modifying cellular regulators of the actin cytoskeleton, namely the Rho GTPases. CNF1 autonomously enters the host cell cytosol, where it catalyzes the constitutive activation of Rho GTPases by deamidation. This activation is, however, attenuated because of activated Rho protein ubiquitin-mediated proteasomal degradation. Both Rho protein activation and deactivation confer phagocytic properties on epithelial and endothelial cells, as well as epithelial cell motility and cell-cell junction dynamics. Transcriptome analysis using DNA microarray revealed that endothelial cells respond to high doses of CNF1 by launching a genetic program of host alarm. This host cell reaction to CNF1 intoxication also indicates that degradation of activated Rho proteins by the proteasome may lead to a lowering of the threshold of the intoxicated cell inflammatory response. These results are consistent with growing evidence that Rho proteins control the cell inflammatory responses. It is tempting to assume that Rho deregulation may participate in various immunological disorders also involved in cancer.

1 Introduction

A fact that has remained a central question for laboratories working on bacterial virulence factors modifying Rho proteins is that, in addition to virulence factors inhibiting Rho proteins, some pathogenic bacteria have evolved virulence factors producing Rho protein activation. What is the rationale for such opposite "strategies" of bacterial virulence? This may represent different requirements for various species of pathogenic bacteria. This idea is illustrated by the findings that Rho protein-inactivating toxins are most frequently encountered in gram-positive bacteria. Nevertheless, this observation does not apply to the observation that Rho protein-activating and -inhibiting virulence factors can both be found in some pathogenic bacteria such as *Salmonellas.*

Rho proteins are key regulators of cellular dynamics and homeostasis, as well as mediators of the cell response to pathogen attack. Given that these aspects on Rho regulation and function are described in other chapters, we will focus this review on recent progress made on CNF1 activity, notably on findings highlighting the hypothesis that pathogenic bacteria, in a "search for optimization" of their interaction with their host, have most likely evolved different mechanisms for moderating their action on host cells, for instance, to prevent them from a massive activation of Rho proteins.

2 The Family of Toxins Activating Rho Proteins

To date, bacterial toxins activating Rho proteins have been isolated exclusively in gram-negative pathogenic bacteria. Cytotoxic necrotizing factors (CNFs) are protein toxins produced by human and animal pathogenic bacteria. CNF1, the first member of the CNFs family described [1], is a chromosomally encoded protein of 1,014 amino acids with a predicted molecular mass of 113.7 kDa. It is found in 30% of uropathogenic *E. coli* isolates [2]. CNF2, isolated from calf and piglet pathogenic *E. coli*, is a plasmid-encoded 110-kDa protein sharing about 90% identity with CNF1 [3]. More recently, Lockman and coworkers have described a CNF-related toxin in *Yersinia pseudotuberculosis*, namely CNFy, which bears 65.1% sequence identity with CNF1 [4]. Finally, a more distant CNF toxin, the dermonecrotic toxin (DNT), has been found in *Bordetella pertussis*, *B. parapertussis*, and *B. bronchiseptica* [5]. DNT is a 160-kDa protein that share sequence homologies with the catalytic domains of CNF1 and CNF2 [5].

The *cnf1* structural gene is located within the pathogenicity island (PAI) PAI-II of the J96 UPEC strain, where it lies between the *hly* and *prs* operons [6]. Unlike *cnfy* and *cnf1*, which are chromosomally encoded, *cnf2* is located on a plasmid [4, 7]. The genetic link between genes encoding CNF1 and α-haemolysin toxins reflects a coregulation in their transcription [8]. In the *E. coli* J96 strain, transcription of *cnf1* is initiated from the *hly* promoter and requires the RfaH antiterminator factor activity [8]. RfaH binds to a specific nucleotide sequence found in the *hly* promoter (*ops/JUMP start*), where it interacts with the RNA polymerase and confers to the enzyme an antiterminator activity [9]. It is established that the *hlyCABD* operon is transcribed in a single mRNA under the positive control of RfaH [10, 11]. *cnf1* transcription thus results from the formation of a polycistronic *hlyCABDcnf1* mRNA [8].

3 How Can CNF1 and DNT Toxins Enter the Cytosol of the Host Cell?

CNF1 is a classic tripartite protein toxin [12]. The amino-terminal third of the toxin contains the cell receptor-binding domain [12]. This allows a tight binding of the toxin to its cognate cell surface receptor, with a K_d as low as 2×10^{-11} M on Hep-2 cells [13]. CNF1 binds the laminin receptor precursor [14]. Binding of the toxin to its cell surface receptor allows its internalization into endocytic vesicles at a low rate and by a nonclathrin endocytosis (independent of Eps15, dynamin, or intersectin-Src homology 3) [13]. CNF1 belongs to a group of bacterial toxins that take advantage of the acidic conditions found in the lumen of endosomes to inject their catalytic domain inside host cell cytosol [13, 15, 16]. By analogy to Diphtheria toxin, it is assumed that, on exposure to acidic conditions, the two hydrophobic helices located in the medium part of the toxin are protonated and inserted into the endosome lipid bilayer, thereby driving the transfer of the catalytic domain to the cytosol [13, 15]. On reaching the cell cytosol, the carboxy-terminal domain of CNF1 catalyzes the posttranslational modification of Rho proteins [17, 18]. Unlike other known bacterial toxins, DNT intoxicates cells by an original mechanism, independent of endosome acidification (bafilomycin insensitive) and of its transfer to the endoplasmic reticulum (brefeldin insensitive) [16, 19]. The binding domain of DNT to its cell surface receptor is encompassed within its 54 N-terminal residues [20]. DNT enters endocytic compartments by a dynamin-dependent endocytosis, where the N1–53 domain is likely cleaved by furin. It is thought that furin cleavage of the amino-terminal 53 residues might unmask hydrophobic helices of the toxin and thus initiates the translocation of the catalytic domain into the cytosol.

4 CNF1 Biochemical Activity: A Permanent Activation of the Rho Proteins

After the discovery that both CNF1 and DNT induced a mobility shift of RhoA on SDS-PAGE, it was hypothesized that these toxins might catalyze a direct posttransductional modification of the GTPase [21, 22]. Soon after, it was discovered through different approaches that CNF1 catalyzed the deamidation of RhoA glutamine 63 into a glutamic acid [17, 18], whereas DNT toxin produced preferentially the transglutamination of RhoA [23]. These biochemical reactions are similar, except that deamidation uses H_2O as acceptor molecules whereas transglutamination uses amine molecules. The specificity of Rho recognition/modification by CNF1 is conferred by RhoA residues R68 and L72, only found at this position in members of the Rho subfamily such as Rho, Rac, and Cdc42 [24]. Consistent with this, Rac and Cdc42 are deamidated by CNF1 in vivo [24, 25]. The three-dimensional structure of the CNF1 catalytic domain revealed that the catalytic triad of CNF1 is buried in a pocket, which most likely confers specificity of Rho accessibility [26]. The glutamine-63 of RhoA modified by CNF1 is a critical amino acid conserved in all known proteins of the Ras superfamily [27]. Glutamine residue 61 of RhoA had long been identified as a hot spot of oncogenic mutation in H-Ras, its mutation being responsible for impairing H-Ras GTPase activity [28]. Similarly, the CNF1-catalyzed deamidation of RhoA glutamine 63 was indeed found to block its GTPase activity, thus conferring permanent Rho activation [17, 18].

5 CNF1-Induced Ubiquitin-Mediated Proteasomal Degradation of Rho Proteins

Studying the extent of Rho protein activation in a model of bladder epithelial cells, to determine the specificity of Rho activation by CNF1, it was observed that cell intoxication resulted in a transient (instead of permanent) activation of Rho proteins [29]. A maximal activation was measured for Rac and Cdc42 isoforms [29]. The transient activation of Rac was directly correlated to the proteasomal degradation of its permanent activated form catalyzed by CNF1 [29, 30] (Fig. 1). Before degradation by the proteasome, cellular proteins are epitope-tagged by conjugation of a polyubiquitin string [31]. Proteasomal degradation of endogenous Rac was found to follow the classic formation of a K48-polyubiquitination chain [29]. Rac sensitivity to ubiquitin-mediated proteasomal degradation appeared to be a direct consequence of the strength of its activation [29]. In addition, it was shown that Rac activation by the GEF

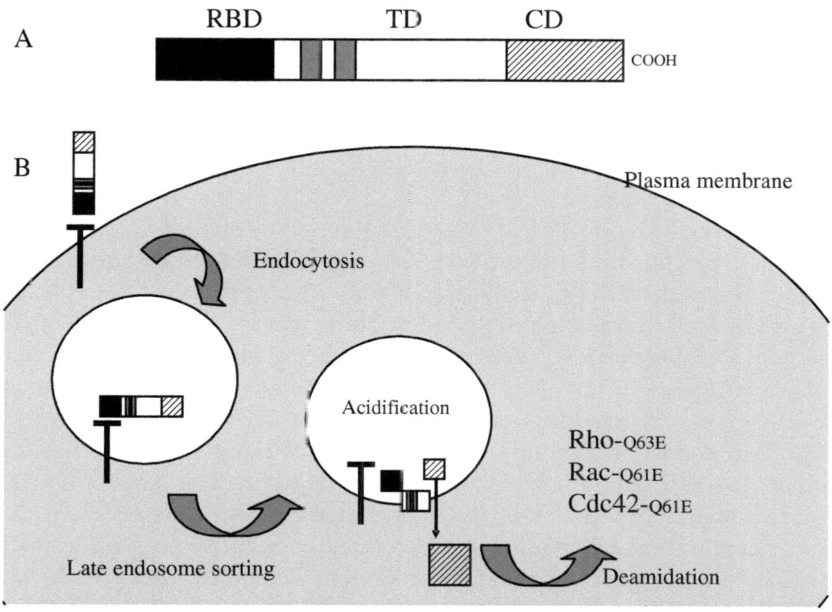

Fig. 1A, B. Structural organization and cell intoxication properties of CNF1. A CNF1 is structured in three functional domains. The amino-terminal part of CNF1 contains the receptor binding domain (*RBD*). The medium part of the toxin contains the translocation domain (*TD*). By analogy to Diphtheria toxin, it is thought that the two hydrophobic helices of this domain have the property of inserting into the lipid bilayer at acidic pH to initiate the transfer of the catalytic carboxy-terminal domain (*CD*) into the cytosol. B Representation of cell intoxication by CNF1. On binding to its cell surface receptor, CNF1 enters into endocytic vesicles. En route to lysosome, CNF1 on reaching a late acidic compartment transfers its catalytic domain into the cell cytosol. Once in the cytosol, the catalytic domain of CNF1 catalyzes the deamidation of the glutamine 63 of RhoA, and its equivalent 61 in Rac or Cdc42, into a glutamic acid. Deamidation of Rho proteins impairs their GTPase activity, conferring them dominant-positive mutant properties. Meanwhile, the activation of Rho proteins sensitizes them to ubiquitin-mediated proteasomal degradation. Together, both Rho protein activation (deamidation) and inactivation (degradation) result in moderating their activation

domain of Dbl also resulted in a significant increase of Rac ubiquitination sensitivity [29]. These results suggested that ubiquitination of Rho proteins might correspond to a yet unraveled regulation of these proteins, which CNF1 may have hijacked. This idea is now sustained by two findings. It was first shown that ubiquitin-mediated proteasomal degradation of RhoB is blocked by TGF-β, resulting in RhoB stabilization [32]. More recently, it was shown that Smad ubiquitination-related factor 1 (Smurf1) bears an ubiquitin ligase activity on RhoA [33]. On the basis of their findings, Wrana and collaborators have hypothesized that Rho ubiquitination might occur during mislocalized activation of RhoA, especially in membrane ruffles where activated RhoA might have antagonized the activity of Rac [33]. Both studies point to a relationship between Rho ubiquitination and TGF-β signaling, which will have to be further clarified.

More complexity arose from recent findings showing, for instance, that DNT activates but does not produce Rac proteasomal degradation in HeLa cells [34] and that CNFy specifically activates RhoA because of the absence of proteasomal degradation of Rho [35]. These reports, together with other findings showing that Rac is activated without being degraded in HEp-2 cells, raise open questions [29]. For instance, these differences might be attributed to cell type specificity. One possibility could be that CNF1 specifically triggers Rho, Rac, and Cdc42 ubiquitin-mediated degradation in epithelial bladder and endothelial cells [29, 36]. These observations may then account for the findings that CNF1 is preferentially encountered in *E. coli* strains responsible for urinary or meningitis infections. Another nonexclusive possibility could be that some cancer cell lines may have a lower ubiquitination activity for one or more Rho protein isoforms. This would raise important questions concerning a possible relationship between ubiquitination of Rho proteins and cancer.

6 Relationship Between Rho Ubiquitin-Mediated Proteasomal Degradation and Host Cells

Rho proteins turned to be a major subject of research since their first description as master transducers of actin cytoskeleton regulation by growth factors [37–40]. On GTP binding, Rho protein members associate with and activate specific effectors [37]. These interaction specificities allow, for instance, Cdc42 and Rac to regulate actin filament assembly producing membrane filopodia or ruffles, respectively [37]. In contrast, Rho controls actin bundling and contraction through regulation of myosin [37]. Through their regulatory properties of actin filament polymerization, organization, and contractility, Rho proteins control a large array of cell processes requiring cell shaping and membrane dynamics [41]. For instance, Rho proteins control the cohesion of cells either between each other or at the contact with the cellular matrix. This aspect is of importance for pathogenic bacteria, whose penetration into host cells is frequently limited because of basolateral localization of their cell internalization receptors [42]. Rho proteins also participate in the control of cell cycle progression and apoptosis [43]. This last aspect is also closely related to pathogen requirements. For instance, it is thought that apoptosis inhibition of target cells may favor bacterial persistence at the epithelium surface and favor bacterial replication and spreading inside host cells [44]. Rho proteins also regulate cell motility and differentiation. These aspects are of particular interest in relation to cellular effectors of the immune response. For instance, macrophage chemotaxis up to the site of bacterial infection is under the control of the actin cytoskeleton machinery, as is bacteria phagocytosis by macrophages [45].

7
CNF1-Triggered Epithelium Invasion

Gram-negative bacteria have thus developed different strategies to penetrate inside cells [42], all of them having in common the use of a family of cellular regulators expressed in all cells, namely, the Rho GTPases. CNF1 produces a counter-intuitive mechanism consisting of Rho protein activation responsible for sensitizing them to ubiquitin-mediated proteasomal degradation [29] (Fig. 2). These observations raise the question of the importance of both Rho protein activation and degradation in bacterial virulence. It has been shown that activation of Rho proteins is necessary to induce CNF1-triggered phagocytosis by epithelial cells [29, 46]. Interestingly, the reaching of a low level of Rho protein activation because of equilibrium between activation and degradation was shown to confer higher invasive properties to pathogenic bacteria (Fig. 2). Similar requirements were also found to confer cell-cell junction dismantling and epithelial cell motility inside monolayers [29]. Urinary tract infections (UTIs) have long been considered as acute and often self-limiting infections caused by noninvasive *E. coli*. Nevertheless, growing evidence suggests that UPECs are individuals capable of colonizing the bladder mucosa. Colonization most likely requires the coordinated action of different virulence factors. At first bacteria attach to bladder epithelial cells through adhesins, preventing then from miction clearing. Persistence of bacteria probably also requires escape from host defenses, comprising innate effectors such as the membrane attack complex of the complement system and immune cell effectors. New evidence suggests that protection of UPECs against host defenses may be achieved by host cell invasion [47]. Taken collectively, these results suggest that CNF1 bears the characteristics of a bona fide invasive factor of intracellular facultative pathogenic bacteria [42]. Intracellular invasion of epithelial cells may not only protect UPECs against host defenses but also may allow bacteria to replicate and/or persist into host cells [44, 48]. Further

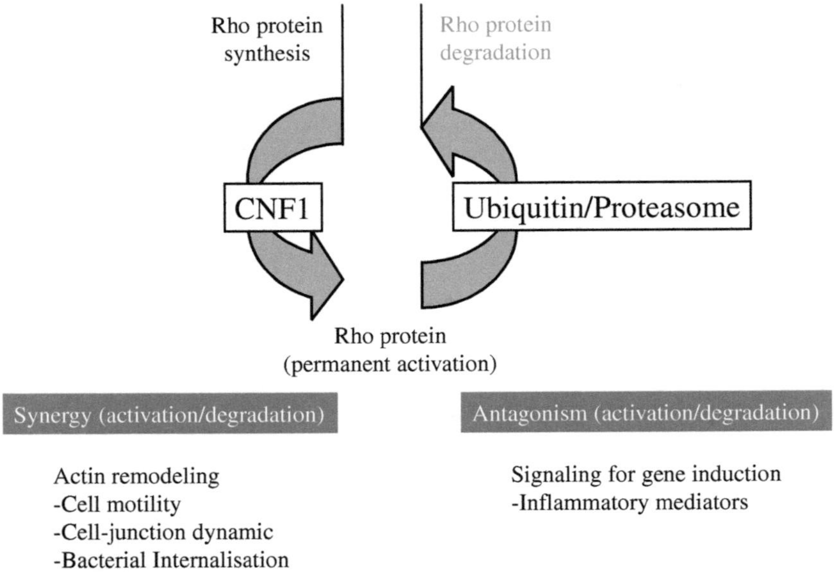

Fig. 2 Host cell response to CNF1-induced Rho activation/degradation. CNF1 produces a counterintuitive mechanism consisting of permanent Rho protein activation followed by Rho sensitization to ubiquitin-mediated proteasomal degradation. This dual mechanism of action confers on host cells efficient bacterial internalization properties, cell-cell junction dynamics, and associated motility inside monolayers. CNF1 activity is thus thought to confer uropathogenic *E. coli* epithelium invasive properties. Cell intoxication by high doses of CNF1 also results in the launching of a cellular program of host alarm and defense. That the level of cytokine production is a direct consequence of the level of Rac/Cdc42 activation suggests that Rho ubiquitin-mediated proteasomal degradation moderates the threshold of host alarm triggered by CNF1

studies will have to clarify whether CNF1 is a major determinant of recurrent cystitis, which infection might then be due to formation of a UPEC reservoir during infection.

8 CNF1-Triggered Gene Response

Studying the transcriptome response of endothelial cells to intoxication by CNF1 revealed that intoxication interferes with classic signaling pathways leading to gene regulation [36]. Consequently, high levels of cell intoxication by CNF1 trigger a gene response consisting of a selective activation of about 0.19% of the 33,000 genes probed on DNA arrays. The 10 most CNF1-activated

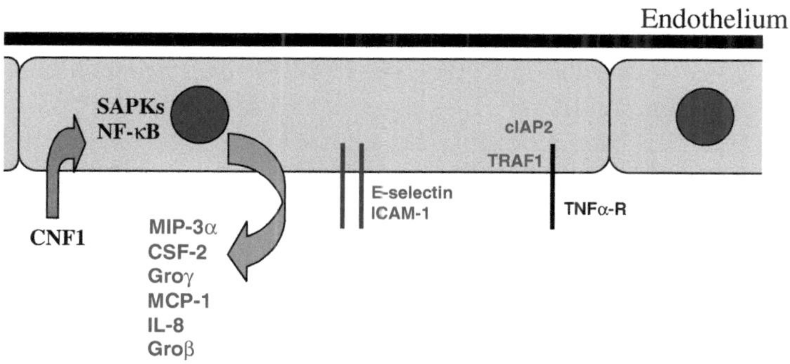

Fig. 3 Transcriptome of the response of endothelial cells to CNF1 intoxication. Human umbilical vein endothelial cell (*HUVEC*) intoxication by CNF1 results in the activation of p38 and c-Jun kinases, stress-activated protein kinase (*SAPK*), as well as NF-κB. Activation of these signaling pathways results in host cells launching a genetic program aimed at leukocyte recruitment and activation. The 10 most CNF1-activated genes are depicted. Induction of inflammatory mediators for leukocyte attraction and activation: *MIP-3α, CSF-2, Groγ, MCP-1, IL-8, Groβ* could be correlated to the production of the leukocyte cell-binding receptors E-selectin (*SELE*) and ICAM-1. Cell response to CNF1 also included TRAF1 (TNF-receptor associated protein-1) and cIAP2 (mammalian inhibitor of apoptosis protein-1 homolog C), two modulators of the TNF-α receptor signaling

genes formed a coherent family of inflammatory mediators, aimed at leukocyte recruitment and activation (Fig. 3). These results are in agreement with a possible effect of CNF1 in the development of UPEC infection, which results in an acute inflammatory disease. The cellular response to CNF1 intoxication includes a large panel of inflammatory effectors responsible for leukocyte recruitment, cell binding, and activation. Induction by CNF1 of membrane metalloproteases and syndecan family products may contribute to the recruitment of leukocytes to the site of infection. Other identified inflammatory regulators such as PLAU-urokinase (for lumen artery restriction), as well as prostaglandin G/H synthesis enzymes, may complete the task of leukocyte recruitment and activation to the site of bacterial infection. Production of GRO-family and MIP-3α chemokines may participate in recruiting lymphocytes and antigen-presenting dendritic cells, respectively. CNF1-intoxicated cells also produce innate defense effectors, comprising complement factor-3 for pathogen phagocytosis and factor-9 (a component of the cytolytic membrane attack complex), as well as GliPR, a plant pathogen-related-1 protein homolog. Finally, the transcriptome of CNF1-intoxicated HUVECs appears to share many similarities with known TNFα-regulated genes, among them TRAF1, cIAP2, and A20. Interestingly, IL8, MCP-1, and MIP-3α production levels were found to be a function of the levels of expression of activated-Cdc42 or Rac. These results also strongly point to CNF1 toxin being an important tool to carry out studies on genes regulated by Rho proteins. Considering that Rac/Cdc42 activation leads to inflammatory mediator production, it is also likely that bacteria have evolved virulence systems aiming at producing a moderate activation of Rho proteins to delay and/or depress the cellular alarm program of the host, while invading cells (Fig. 2).

The moderate activation of Rho proteins may be beneficial during early stages of the bacterial infection, prior substantive bacterial growth, and a resulting acute inflammatory reaction due to, for instance, pathogen-associated molecular pattern recognition (PAMP) by Toll-like receptors (TLRs) [49]. In that respect, the mechanism of action of CNF1 consisting in Rho protein activation (deamidation) and deactivation (ubiquitin-mediated proteasomal degradation) may be sharply adapted to produce a moderate activation of Rho proteins. This idea is illustrated by the findings that TLRs can activate Rac and Cdc42, this last GTPase being required for TLR signal transduction [50]. Together these observations raise the question of why bacterial virulence factors activate a transduction pathway of host alarm and defense.

A similar question also arose from the findings that commensal bacteria are recognized by TLRs and that this recognition triggers a production of IL6, TNF, and KC-1 required for gut protection [51]. In fact, this work revealed that TLRs control intestinal epithelial homeostasis and protection against injury. It also provides a rational explanation for the findings that prokaryotes have evolved virulence factors to inhibit the $SCF^{\beta\text{-TrCP}}$ ubiquitination of Iκ-Bα, a signal required for proper TLR signal transduction [52]. Similarly, it is striking to observe that the CNF1-activated genes in HUVECs are similar to those triggered on HUVEC binding to fibronectin, a matrix protein that results in Rac activation [53, 54]. In this latter condition it is assumed that HUVECs associate to fibronectin on wounding and that this cellular response corresponds to the inflammatory reaction needed for tissue repair. All of these studies reinforce the idea that cells respond similarly to different environmental injuries, among them pathogen attack. These findings thus raise important questions of the nature of the inflammatory responses triggered by Rho protein activation. The fact that Rho proteins appeared to be key regulators of the cell immune responses raises important questions on their participation in various inflammatory disorders comprising immune escape of cancer cells [55].

References

1. Falbo V, Pace T, Picci L, Pizzi E & Caprioli A (1993) Isolation and nucleotide sequence of the gene encoding cytotoxic necrotizing factor 1 of *Escherichia coli.* Infect Immun. 61:4909–4914.
2. Landraud L, Gauthier M, Fosse T & Boquet P (2000) Frequency of *Escherichia coli* strains producing the cytotoxic necrotizing factor (CNF1) in nosocomial urinary tract infections. Lett. Appl. Microbiol. 30:213–216.
3. De Rycke J, Gonzalez EA, Blanco J, Oswald E, Blanco M & Boivin R (1990) Evidence for two types of cytotoxic necrotizing factor in human and animal clinical isolates of *Escherichia coli.* J. Clin. Microbiol. 28:694–699.
4. Lockman HA, Gillespie RA, Baker BD & Shakhnovich E (2002) *Yersinia pseudotuberculosis* produces a cytotoxic necrotizing factor. Infect. Immun. 70:2708–2714.
5. Horiguchi Y (2001) *Escherichia coli* cytotoxic necrotizing factors and *Bordetella* dermonecrotic toxin: the dermonecrosis-inducing toxins activating Rho small GTPases. Toxicon 39:1619–1627.
6. Swenson DL, Bukanov NO, Berg DE & Welch RA (1996) Two pathogenicity islands in uropathogenic *Escherichia coli* J96: cosmid cloning and sample sequencing. Infect Immun. 64:3736–3743.
7. Oswald E & De Rycke J (1990) A single protein of 110 kDa is associated with the multinucleating and necrotizing activity coded by the Vir plasmid of *Escherichia coli.* FEMS Microbiol Lett. 56:279–284.
8. Landraud L, Gibert M, Popoff MR, Boquet P & Gauthier M (2003) Expression of cnf1 by *Escherichia coli* J96 involves a large upstream DNA region including the hlyCABD operon, and is regulated by the RfaH protein. Mol. Microbiol. 47:1653–1667.
9. Artsimovitch I & Landick R (2002) The transcriptional regulator RfaH stimulates RNA chain synthesis after recruitment to elongation complexes by the exposed nontemplate DNA strand. Cell 109:193–203.
9. Leeds JA & Welch RA (1997) Enhancing transcription through the *Escherichia coli* hemolysin operon, hlyCABD: RfaH and upstream JUMPStart DNA sequences function together via a postinitiation mechanism. J Bacteriol 179:3519–3527.
10. Bailey MJ, Hughes C & Koronakis V (1997) RfaH and the ops element, components of a novel system controlling bacterial transcription elongation. Mol. Microbiol. 26:845–851.
12. Lemichez E, Flatau G, Bruzzone M, Boquet P & Gauthier M (1997) Molecular localization of the *Escherichia coli* cytotoxic necrotizing factor CNF1 cell-binding and catalytic domains. Mol. Microbiol. 24:1061–1070.
13. Contamin S, Galmiche A, Doye A, Flatau G, Benmerah A & Boquet P (2000) The p21 Rho-activating toxin cytotoxic necrotizing factor 1 is endocytosed by a clathrin-independent mechanism and enters the cytosol by an acidic-dependent membrane translocation step. Mol. Biol. Cell 11:1775–1787.
14. Chung JW, Hong SJ, Kim KJ, Goti D, Stins MF, Shin S, Dawson VL, Dawson TM & Kim KS (2003) 37-kDa laminin receptor precursor modulates cytotoxic necrotizing factor 1-mediated RhoA activation and bacterial uptake. J. Biol. Chem. 278:16857–16862.
15. Pei S, Doye A & Boquet P (2001) Mutation of specific acidic residues of the CNF1 T domain into lysine alters cell membrane translocation of the toxin. Mol. Microbiol. 41:1237–1247.

16. Lemichez E & Boquet P (2003) To be helped or not helped, that is the question. J. Cell Biol. 160:991–992.
17. Flatau G, Lemichez E, Gauthier M, Chardin P, Paris S, Fiorentini C & Boquet P (1997) Toxin-induced activation of the G protein p21 Rho by deamidation of glutamine. Nature 387:729–733.
18. Schmidt G, Sehr P, Wilm M, Selzer J, Mann M & Aktories K (1997) Gln 63 of Rho is deamidated by *Escherichia coli* cytotoxic necrotizing factor-1. Nature 387:725–729.
19. Matsuzawa T, Fukui A, Kashimoto T, Nagao K, Oka K, Miyake M & Horiguchi Y (2004) *Bordetella* dermonecrotic toxin undergoes proteolytic processing to be translocated from a dynamin-related endosome into the cytoplasm in an acidification-independent manner. J Biol Chem. 279:2866–2872.
20. Matsuzawa T, Kashimoto T, Katahira J & Horiguchi Y (2002) Identification of a receptor-binding domain of *Bordetella* dermonecrotic toxin. Infect. Immun. 70:3427–3432.
21. Oswald E, Sugai M, Labigne A, Wu HC, Fiorentini C, Boquet P & O'Brien AD (1994) Cytotoxic necrotizing factor type 2 produced by virulent *Escherichia coli* modifies the small GTP-binding proteins Rho involved in assembly of actin stress fibers. Proc. Natl. Acad. Sci. USA. 91:3814–388.
22. Horiguchi Y, Senda T, Sugimoto N, Katahira J & Matsuda M (1995) *Bordetella bronchiseptica* dermonecrotizing toxin stimulates assembly of actin stress fibers and focal adhesions by modifying the small GTP-binding protein rho. J. Cell Sci. 108:3243–3251.
23. Masuda M, Betancourt L, Matsuzawa T, Kashimoto T, Takao T, Shimonishi Y & Horiguchi Y (2000) Activation of rho through a cross-link with polyamines catalyzed by *Bordetella* dermonecrotizing toxin. EMBO J. 19:521–530.
24. Lerm M, Schmidt G, Goehring UM, Schirmer J & Aktories K (1999) Identification of the region of rho involved in substrate recognition by *Escherichia coli* cytotoxic necrotizing factor 1 (CNF1). J. Biol. Chem. 274:28999–29004.
25. Lerm M, Selzer J, Hoffmeyer A, Rapp UR, Aktories K & Schmidt G (1999) Deamidation of Cdc42 and Rac by *Escherichia coli* cytotoxic necrotizing factor 1: activation of c-Jun N-terminal kinase in HeLa cells. Infect. Immun. 67:496-503.
26. Buetow L, Flatau G, Chiu K, Boquet P & Ghosh P (2001) Structure of the Rho-activating domain of *Escherichia coli* cytotoxic necrotizing factor 1. Nat. Struct. Biol. 8:584–588.
27. Takai Y, Sasaki T & Matozaki T (2001) Small GTP-binding proteins. Physiol. Rev. 81:153–208.
28. Der CJ, Finkel T & Cooper GM (1986) Biological and biochemical properties of human rasH genes mutated at codon 61. Cell 44:167–176.
29. Doye A et al. (2002) CNF1 exploits the ubiquitin-proteasome machinery to restrict Rho GTPase activation for bacterial host cell invasion. Cell 111:553–564.
30. Lerm M, Pop M, Fritz G, Aktories K & Schmidt G (2002) Proteasomal degradation of cytotoxic necrotizing factor 1-activated rac. Infect. Immun. 70:4053–4058.
31. Finley D, Ciechanover A & Varshavsky A (2004) Ubiquitin as a central cellular regulator. Cell 116:S29–S32.
32. Engel ME, Datta PK & Moses HL (1998) RhoB is stabilized by transforming growth factor $\beta$and antagonizes transcriptional activation. J. Biol. Chem. 273:9921–9926.
33. Wang HR et al. (2003) Regulation of cell polarity and protrusion formation by targeting RhoA for degradation. Science 302:1775–1779.
34. Hoffmann C, Pop M, Leemhuis J, Schirmer J, Aktories K & Schmidt G (2004) The *Yersinia pseudotuberculosis* cytotoxic necrotizing factor (CNFY) selectively activates RhoA. J. Biol. Chem. 279:16026–16032.

35. Pop M, Aktories K & Schmidt G (2004) Isotype-specific degradation of Rac activated by the cytotoxic necrotizing factor 1. J. Biol. Chem. 279:35840–3588.
36. Munro P, Flatau G, Doye A, Boyer L, Oregioni O, Mege JL, Landraud L & Lemichez E (2004) Activation and proteasomal degradation of rho GTPases by cytotoxic necrotizing factor-1 elicit a controlled inflammatory response. J. Biol. Chem. 279:35849–35857.
37. Burridge K & Wennerberg K (2004) Rho and Rac take center stage. Cell 116:167–79.
38. Ridley A J & Hall A (1992) The small GTP-binding protein rho regulates the assembly of focal adhesions and actin stress fibers in response to growth factors. Cell 70:389–399.
39. Ridley AJ, Paterson HF, Johnston CL, Diekmann D & Hall, A (1992) The small GTP-binding protein rac regulates growth factor-induced membrane ruffling. Cell 70:401–410.
40. Nobes CD & Hall A (1995) Rho, rac, and cdc42 GTPases regulate the assembly of multimolecular focal complexes associated with actin stress fibers, lamellipodia, and filopodia. Cell 81:53–62.
41. Etienne-Manneville S & Hall A (2002) Rho GTPases in cell biology. Nature 420:629–635.
42. Cossart P & Sansonetti PJ (2004) Bacterial invasion: the paradigms of enteroinvasive pathogens. Science 304:242–248.
43. Sahai E & Marshall CJ (2002) Rho-GTPases and cancer. Nat. Rev. Cancer 21:133–142.
44. Mulvey MA, Lopez-Boado YS, Wilson CL, Roth R, Parks WC, Heuser J & Hultgren SJ (1998) Induction and evasion of host defenses by type 1-piliated uropathogenic *Escherichia coli*. Science 282:1494–1497.
45. Caron E & Hall A (1998) Identification of two distinct mechanisms of phagocytosis controlled by different Rho GTPases. Science 282:1717–1721.
46. Falzano L, Fiorentini C, Donelli G, Michel E, Kocks C, Cossart P, Cabanie L & Oswald E & Boquet P (1993) Induction of phagocytic behaviour in human epithelial cells by *Escherichia coli* cytotoxic necrotizing factor type 1. Mol. Microbiol. 9:1247–1254.
47. Schilling JD, Mulvey MA & Hultgren SJ (2001) Dynamic interactions between host and pathogen during acute urinary tract infections. Urology 57:56–61.
48. Svanborg C, Godaly G & Hedlund M (1999) Cytokine responses during mucosal infections: role in disease pathogenesis and host defense. Curr. Opin. Microbiol. 2:99–105.
49. Janssens S & Beyaert R (2003) Role of Toll-like receptors in pathogen recognition. Clin. Microbiol. Rev. 16:637–646.
50. Arbibe L, Mira JP, Teusch N, Kline L, Guha M, Mackman N, Godowski PJ, Ulevitch RJ & Knaus UG (2000) Toll-like receptor 2-mediated NF-κB activation requires a Rac1-dependent pathway. Nat. Immunol. 1:533–540.
51. Rakoff-Nahoum S, Paglino J, Eslami-Varzaneh F, Edberg S & Medzhitov R (2004) Recognition of commensal microflora by toll-like receptors is required for intestinal homeostasis. Cell 118:229–241.
52. Neish AS, Gewirtz AT, Zeng H, Young AN, Hobert ME, Karmali V, Rao AS & Madara JL (2000) Prokaryotic regulation of epithelial responses by inhibition of IκB-α ubiquitination. Science. 289:1560–1563.
53. Mettouchi A, Klein S, Guo W, Lopez-Lago M, Lemichez E, Westwick JK & Giancotti FG (2001) Integrin-specific activation of Rac controls progression through the G_1 phase of the cell cycle. Mol. Cell 8:115–127.

54. Klein S, de Fougerolles AR, Blaikie P, Khan L, Pepe A, Green CD, Koteliansky V & Giancotti FG (2002) $\alpha_5\beta_1$ Integrin activates an NF-κB-dependent program of gene expression important for angiogenesis and inflammation. Mol. Cell Biol. 22:5912–5922.
55. Boettner B & Van Aelst L (2002) The role of Rho GTPases in disease development. Gene 286:155–74.

Subject Index

Current Topics in Microbiology and Immunology

Volumes published since 1989 (and still available)

Vol. 246: **Melchers, Fritz; Potter, Michael (Eds.):** Mechanisms of B Cell Neoplasia 1998. 1999. 111 figs. XXIX, 415 pp. ISBN 3-540-65759-2

Vol. 247: **Wagner, Hermann (Ed.):** Immunobiology of Bacterial CpG-DNA. 2000. 34 figs. IX, 246 pp. ISBN 3-540-66400-9

Vol. 248: **du Pasquier, Louis; Litman, Gary W. (Eds.):** Origin and Evolution of the Vertebrate Immune System. 2000. 81 figs. IX, 324 pp. ISBN 3-540-66414-9

Vol. 249: **Jones, Peter A.; Vogt, Peter K. (Eds.):** DNA Methylation and Cancer. 2000. 16 figs. IX, 169 pp. ISBN 3-540-66608-7

Vol. 250: **Aktories, Klaus; Wilkins, Tracy, D. (Eds.):** Clostridium difficile. 2000. 20 figs. IX, 143 pp. ISBN 3-540-67291-5

Vol. 251: **Melchers, Fritz (Ed.):** Lymphoid Organogenesis. 2000. 62 figs. XII, 215 pp. ISBN 3-540-67569-8

Vol. 252: **Potter, Michael; Melchers, Fritz (Eds.):** B1 Lymphocytes in B Cell Neoplasia. 2000. XIII, 326 pp. ISBN 3-540-67567-1

Vol. 253: **Gosztonyi, Georg (Ed.):** The Mechanisms of Neuronal Damage in Virus Infections of the Nervous System. 2001. approx. XVI, 270 pp. ISBN 3-540-67617-1

Vol. 254: **Privalsky, Martin L. (Ed.):** Transcriptional Corepressors. 2001. 25 figs. XIV, 190 pp. ISBN 3-540-67569-8

Vol. 255: **Hirai, Kanji (Ed.):** Marek's Disease. 2001. 22 figs. XII, 294 pp. ISBN 3-540-67798-4

Vol. 256: **Schmaljohn, Connie S.; Nichol, Stuart T. (Eds.):** Hantaviruses. 2001, 24 figs. XI, 196 pp. ISBN 3-540-41045-7

Vol. 257: **van der Goot, Gisou (Ed.):** Pore-Forming Toxins, 2001. 19 figs. IX, 166 pp. ISBN 3-540-41386-3

Vol. 258: **Takada, Kenzo (Ed.):** Epstein-Barr Virus and Human Cancer. 2001. 38 figs. IX, 233 pp. ISBN 3-540-41506-8

Vol. 259: **Hauber, Joachim, Vogt, Peter K. (Eds.):** Nuclear Export of Viral RNAs. 2001. 19 figs. IX, 131 pp. ISBN 3-540-41278-6

Vol. 260: **Burton, Didier R. (Ed.):** Antibodies in Viral Infection. 2001. 51 figs. IX, 309 pp. ISBN 3-540-41611-0

Vol. 261: **Trono, Didier (Ed.):** Lentiviral Vectors. 2002. 32 figs. X, 258 pp. ISBN 3-540-42190-4

Vol. 262: **Oldstone, Michael B.A. (Ed.):** Arenaviruses I. 2002, 30 figs. XVIII, 197 pp. ISBN 3-540-42244-7

Vol. 263: **Oldstone, Michael B. A. (Ed.):** Arenaviruses II. 2002, 49 figs. XVIII, 268 pp. ISBN 3-540-42705-8

Vol. 264/I: **Hacker, Jörg; Kaper, James B. (Eds.):** Pathogenicity Islands and the Evolution of Microbes. 2002. 34 figs. XVIII, 232 pp. ISBN 3-540-42681-7

Vol. 264/II: **Hacker, Jörg; Kaper, James B. (Eds.):** Pathogenicity Islands and the Evolution of Microbes. 2002. 24 figs. XVIII, 228 pp. ISBN 3-540-42682-5

Vol. 265: **Dietzschold, Bernhard; Richt, Jürgen A. (Eds.):** Protective and Pathological Immune Responses in the CNS. 2002. 21 figs. X, 278 pp. ISBN 3-540-42668-X

Vol. 266: **Cooper, Koproski (Eds.):** The Interface Between Innate and Acquired Immunity, 2002, 15 figs. XIV, 116 pp. ISBN 3-540-42894-X

Vol. 267: **Mackenzie, John S.; Barrett, Alan D. T.; Deubel, Vincent (Eds.):** Japanese Encephalitis and West Nile Viruses. 2002. 66 figs. X, 418 pp. ISBN 3-540-42783X

Vol. 268: **Zwickl, Peter; Baumeister, Wolfgang (Eds.):** The Proteasome-Ubiquitin Protein Degradation Pathway. 2002, 17 figs. X, 213 pp.
ISBN 3-540-43096-2

Vol. 269: **Koszinowski, Ulrich H.; Hengel, Hartmut (Eds.):** Viral Proteins Counteracting Host Defenses. 2002, 47 figs. XII, 325 pp. ISBN 3-540-43261-2

Vol. 270: **Beutler, Bruce; Wagner, Hermann (Eds.):** Toll-Like Receptor Family Members and Their Ligands. 2002, 31 figs. X, 192 pp. ISBN 3-540-43560-3

Vol. 271: **Koehler, Theresa M. (Ed.):** Anthrax. 2002, 14 figs. X, 169 pp.
ISBN 3-540-43497-6

Vol. 272: **Doerfler, Walter; Böhm, Petra (Eds.):** Adenoviruses: Model and Vectors in Virus-Host Interactions. Virion and Structure, Viral Replication, Host Cell Interactions. 2003, 63 figs., approx. 280 pp.
ISBN 3-540-00154-9

Vol. 273: **Doerfler, Walter; Böhm, Petra (Eds.):** Adenoviruses: Model and Vectors in Virus-Host Interactions. Immune System, Oncogenesis, Gene Therapy. 2004, 35 figs., approx. 280 pp. ISBN 3-540-06851-1

Vol. 274: **Workman, Jerry L. (Ed.):** Protein Complexes that Modify Chromatin. 2003, 38 figs., XII, 296 pp. ISBN 3-540-44208-1

Vol. 275: **Fan, Hung (Ed.):** Jaagsiekte Sheep Retrovirus and Lung Cancer. 2003, 63 figs., XII, 252 pp. ISBN 3-540-44096-3

Vol. 276: **Steinkasserer, Alexander (Ed.):** Dendritic Cells and Virus Infection. 2003, 24 figs., X, 296 pp. ISBN 3-540-44290-1

Vol. 277: **Rethwilm, Axel (Ed.):** Foamy Viruses. 2003, 40 figs., X, 214 pp.
ISBN 3-540-44388-6

Vol. 278: **Salomon, Daniel R.; Wilson, Carolyn (Eds.):** Xenotransplantation. 2003, 22 figs., IX, 254 pp.
ISBN 3-540-00210-3

Vol. 279: **Thomas, George; Sabatini, David; Hall, Michael N. (Eds.):** TOR. 2004, 49 figs., X, 364 pp.
ISBN 3-540-00534-X

Vol. 280: **Heber-Katz, Ellen (Ed.):** Regeneration: Stem Cells and Beyond. 2004, 42 figs., XII, 194 pp. ISBN 3-540-02238-4

Vol. 281: **Young, John A. T. (Ed.):** Cellular Factors Involved in Early Steps of Retroviral Replication. 2003, 21 figs., IX, 240 pp. ISBN 3-540-00844-6

Vol. 282: **Stenmark, Harald (Ed.):** Phosphoinositides in Subcellular Targeting and Enzyme Activation. 2003, 20 figs., X, 210 pp. ISBN 3-540-00950-7

Vol. 283: **Kawaoka, Yoshihiro (Ed.):** Biology of Negative Strand RNA Viruses: The Power of Reverse Genetics. 2004, 24 figs., IX, 350 pp. ISBN 3-540-40661-1

Vol. 284: **Harris, David (Ed.):** Mad Cow Disease and Related Spongiform Encephalopathies. 2004, 34 figs., IX, 219 pp. ISBN 3-540-20107-6

Vol. 285: **Marsh, Mark (Ed.):** Membrane Trafficking in Viral Replication. 2004, 19 figs., IX, 259 pp. ISBN 3-540-21430-5

Vol. 286: **Madshus, Inger H. (Ed.):** Signalling from Internalized Growth Factor Receptors. 2004, 19 figs., IX, 187 pp.
ISBN 3-540-21038-5

Vol. 287: **Enjuanes, Luis (Ed.):** Coronavirus Replication and Reverse Genetics. 2005, 49 figs., XI, 257 pp. ISBN 3-540-21494-1

Vol. 288: **Mahy, Brain W. J. (Ed.):** Foot-and-Mouth-Disease Virus. 2005, 16 figs., IX, 178 pp. ISBN 3-540-22419-X

Vol. 289: **Griffin, Diane E. (Ed.):** Role of Apoptosis in Infection. 2005, 40 figs., IX, 294 pp. ISBN 3-540-23006-8

Vol. 290: **Singh, Harinder; Grosschedl, Rudolf (Eds.):** Molecular Analysis of B Lymphocyte Development and Activation. 2005, 28 figs., XI, 255 pp.
ISBN 3-540-23090-4

MIX
Papier aus verantwortungsvollen Quellen
Paper from responsible sources
FSC® C105338

If you have any concerns about our products,
you can contact us on
ProductSafety@springernature.com

In case Publisher is established outside the EU,
the EU authorized representative is:
Springer Nature Customer Service Center GmbH
Europaplatz 3, 69115 Heidelberg, Germany

Printed by Libri Plureos GmbH
in Hamburg, Germany